VIDEO TRACKING

VIDEO TRACKING
Theory and Practice

Emilio Maggio
Vicon, UK

Andrea Cavallaro
Queen Mary University of London, UK

WILEY

A John Wiley and Sons, Ltd., Publication

This edition first published 2011
© 2011, John Wiley & Sons, Ltd

Registered office
John Wiley & Sons Ltd, The Atrium, Southern Gate, Chichester, West Sussex, PO19
8SQ, United Kingdom

For details of our global editorial offices, for customer services and for information about
how to apply for permission to reuse the copyright material in this book please see our
website at www.wiley.com.

Library of Congress Cataloging-in-Publication Data

Cavallaro, Andrea.
    Video tracking : theory and practice / Andrea Cavallaro, Emilio Maggio.
      p. cm.
    Includes bibliographical references and index.
    ISBN 978-0-470-74964-7 (cloth)
1. Video surveillance.   2. Automatic tracking.   I. Maggio, Emilio.   II. Title.
    TK6680.3.C38 2010
    621.389′28–dc22

                                                              2010026296

A catalogue record for this book is available from the British Library.

Print ISBN: 978-0-4707-4964-7
ePDF ISBN: 978-0-4709-7438-4
oBook ISBN: 978-0-4709-7437-7

Set in 10/12pt cmr by Aptara Inc., New Delhi, India.

CONTENTS

v

FOREWORD

I am honored to have been asked to write a foreword to this comprehensive, timely and extremely well-written book on Video Tracking: Theory and Practice by Prof. Andrea Cavallaro and Dr. Emilio Maggio. The book is comprehensive in that it brings together theory and methods developed since the early sixties for point object tracking that dominated aerospace applications and the so called extended object tracking that arises in computer vision and image processing applications. The publication of this book is timely as it provides a one stop source for learning about the voluminous body of literature on video tracking that has been generated in the last fifteen years or so. The book written in a lucid style will help students, researchers and practicing engineers to quickly learn about what has been done and what needs to be done in this important field. The field of computer vision is populated by computer scientists and electrical engineers who often have different levels of familiarity with principles of random process, detection and estimation theory. This book is written in a way such that it is easily accessible to computer scientists, not all of who have taken graduate level courses in random process and estimation theory, while at the same time interesting enough to electrical engineers who may be aware of the underlying theory but are not cognizant of the myriad of applications.

Early work on video tracking was mostly concerned with tracking of point objects using infrared sensors with military applications. The alpha-beta tracker developed in the early years of this field soon gave way to the magic of Kalman filter (continuous and discrete) and its variants. It is not an exaggeration to say that most existing systems for tracking objects are built on some versions of the Kalman filter. Discussions of Kalman-filter based trackers may be found in the many books written by Anderson and Moore, Bar-Shalom and colleagues, Gelb, Blackman and Popli and many others. Theoretical underpinnings for the design of trackers are discussed in the classical book by Jaswinski. Linear Kalman filters have been phenomenally effective for tracking problems that can be modeled using linear systems corrupted by Gaussian noise. Their effectiveness for non-linear and/or non-Gaussian tracking problems has been mixed; the design of extended Kalman filters, iterated extended Kalman filters and non-linear continuous trackers for non-linear/non-Gaussian tracking problems appears to be guided by science and art!

When trackers built for tracking point objects are generalized to tracking extended objects like faces, humans and vehicles, several challenges have to be addressed. These challenges include addressing the variations due to geometric (pose, articulation) and photometric factors (illumination, appearance). Thus one needs to either extract features that are invariant to these factors and use them for tracking or incorporate these variations in an explicit way in the design of trackers. The need to incorporate geometric and photometric variations has necessitated the development of more sophisticated trackers based on Monte Carlo Markov Chain techniques; the popular particle filters belong to this family of trackers. Recently, several research monographs on the theory and applications of particle filters to video tracking have appeared in the literature. These are cited in the chapter on further reading.

This book begins with an inviting chapter that introduces the topic of video tracking and challenges in designing a robust video tracker and then presents an easy to follow outline of the book. Mathematical notations and simple formulations of single and multi-object trackers are then given. It is not an exaggeration to say that tracking is an application-driven problem. Chapter 2 presents an excellent overview of applications from entertainment, healthcare, surveillance, and robotics. Recent applications to object tracking using sensors mounted on unmanned platforms are also discussed. Chapter 3 gives a nice summary of numerous features (intensity, color, gradients, regions of interest and even object models) that one can extract for video tracking and their relative robustness to the variations mentioned above. The chapter also presents related material on image formation and preprocessing algorithms for background subtraction.

Chapter 4 gives an excellent summary of models for shape, deformations and appearance. Methods for coping with variations of these representations are also discussed. Representation of tracked objects (appearance and motion) is critical in defining the state and measurement equations for designing Kalman filters or for deriving the probabilistic models for designing particle filters. Chapter 5 on tracking algorithms for single objects can be considered as the "brain" of this book. Details of theory and implementations of Kalman filters and particle filters are given at a level that is at once appealing to a wide segment of population.

Fusion of multiple attributes has been historically noted to improve the performance of trackers. In video tracking applications, fusing motion and intensity-based attributes helps with tracking dim targets in cluttered environments. Chapter 6 presents the basics of fusion methodologies that will help the design of shape, motion, behavior and other attributes for realizing robust trackers. When multiple objects have to be tracked, associating them is the central problem to be tackled. Popular strategies for probabilistic data association that have been around for more than two decades along with more recently developed methods based on graph theory are elegantly discussed in Chapter 7. When tracking multiple objects or multiple features on a moving object, it is likely that some features will disappear and new features will arise. This chapter also presents for handling the *birth* and *death* of features; this is very important for long-duration tracking. The authors should be congratulated for having presented this difficult topic in a very easy to read manner.

The notion of incorporating context in object detection, recognition and tracking has engaged the minds of computer vision researchers since the early nineties. Chapter 8 discusses the role of context in improving the performance of video trackers. Methods for extracting contextual information (to determine where an object may be potentially found) which can be incorporated in the trackers are also discussed. One of the practical aspects of designing trackers is to be able to provide some performance bounds on how well the trackers work. Many workshops have been organized for discussing the vexing but important problem of performance evaluation of trackers using metrics, common data bases etc. Although sporadic theoretical analyses has been done for evaluating the performance of trackers, most of the existing methods are empirical based on ground truthed data. Chapter 9 is a must read chapter to understand the history and practice of how trackers are evaluated.

The book ends with an epilogue that briefly discusses future challenges that need to be addressed, a strong appendix on comparison of several trackers discussed in the book and suggestions for further reading.

I enjoyed reading this book which brings close to three decades of work on video tracking in a single book. The authors have taken into consideration the needs and backgrounds of the potential readers from image processing and computer vision communities and have written a book that will help the researchers, students and practicing engineers to enter and stay in this important field. They have not only created a scholarly account of the science produced by most researchers but have also emphasized the applications, keeping in mind the science, art and technology of the field.

Rama Chellappa
College Park, Maryland.

ABOUT THE AUTHORS

Emilio Maggio is a Computer Vision Scientist at Vicon, the motion capture worldwide market leader. His research is on object tracking, classification, Bayesian filtering, sparse image and video coding. In 2007 he was a visiting researcher at Mitsubishi Research Labs (MERL) and in 2003 he visited the Signal Processing Institute at the Swiss Federal Institute of Technology (EPFL). Dr. Maggio has been twice awarded a best student paper prize at IEEE ICASSP, in 2005 and 2007; he also won in 2002, the IEEE Computer Society International Design Competition.

Andrea Cavallaro is Professor of Multimedia Signal Processing at Queen Mary University of London. His research is on target tracking and multimodal content analysis for multi-sensor systems. He was awarded a Research Fellowship with BT labs in 2004, the Royal Academy of Engineering teaching Prize in 2007; three student paper awards at IEEE ICASSP in 2005, 2007 and 2009; and the best paper award at IEEE AVSS 2009. Dr. Cavallaro is Associate Editor for the IEEE Signal Processing Magazine, the IEEE Transactions on Multimedia and the IEEE Transactions on Signal Processing.

PREFACE

Video tracking is the task of estimating over time the position of objects of interest in image sequences. This book is the first offering a comprehensive and dedicated coverage of the emerging topic of video tracking, the fundamental aspects of algorithm development and its applications. The book introduces, discusses and demonstrates the latest video-tracking algorithms with a unified and comprehensive coverage.

Starting from the general problem definition and a review of existing and emerging applications, we introduce popular video trackers, such as those based on correlation and gradient-descent minimisation. We discuss, using practical examples and illustrations as support, the advantages and limitations of deterministic approaches and then we promote the use of more efficient and accurate video-tracking solutions. Recent algorithms based on the Bayes' recursive framework are presented and their application to real-word tracking scenarios is discussed. Throughout the book we discuss the design choices and the implementation issues that are necessary to turn the underlying mathematical modelling into a real-world effective system. To facilitate learning, the book provides block diagrams and simil-code implementations of the algorithms.

Chapter 1 introduces the video-tracking problem and presents it in a unified view by dividing the problem into five main logical tasks. Next, the chapter provides a formal problem formulation for video tracking. Finally, it discusses typical challenges that make video tracking difficult. *Chapter 2* discusses current and emerging applications of video tracking. Application areas include media production, medical data processing, surveillance, business intelligence, robotics, tele-collaboration, interactive gaming and art.

Chapter 3 offers a high-level overview of the video acquisition process and presents relevant features that can be selected for the representation of a target. Next, *Chapter 4* discusses various shape-approximation strategies and appearance modelling techniques.

Chapter 5 introduces a taxonomy for localisation algorithms and compares single and multi-hypothesis strategies. *Chapter 6* discusses the modalities for fusing multiple features for target tracking. Advantages and disadvantages of fusion at the tracker level and at the feature level are discussed. Moreover, we present appropriate measures for quantifying the reliability of features prior to their combination.

Chapter 7 extends the concepts covered in the first part of the book to tracking a variable number of objects. To better exemplify these methods, particular attention is given to multi-hypothesis data-association algorithms applied to video surveillance. Moreover, the chapter discusses and evaluates the first video-based multi-target tracker based on finite set statistics. Using this tracker as an example, *Chapter 8* discusses how modelling the scene can help improve the performance of a video tracker. In particular, we discuss automatic and interactive strategies for learning areas of interest in the image.

Chapter 9 covers protocols to be used to formally evaluate a video tracker and the results it generates. The chapter provides the reader with a comprehensive overview of performance measures and a range of evaluation datasets.

Finally, the *Epilogue* summarises the current directions and future challenges in video tracking and offers a further reading list, while the *Appendix* reports and discusses comparative numerical results of selected methods presented in the book.

The book is aimed at graduate students, researchers and practitioners interested in the various vision-based interpretive applications, smart environments, behavioural modelling, robotics and video annotation, as well as application developers in the areas of surveillance, motion capture, virtual reality and medical-image sequence analysis.

The website of the book, www.videotracking.org, includes a comprehensive list of software algorithms that are publicly available for video tracking and offers to instructors for use in the classroom a series of PowerPoint presentations covering the material presented in the book.

Emilio Maggio and Andrea Cavallaro
London, UK

ACKNOWLEDGEMENTS

We would like to acknowledge the contribution of several people who helped make this book a reality. The book has grown out of more than a decade of thinking, experimenting and teaching. Along the way we have become indebted to many colleagues, students and friends. Working with them has been a memorable experience. In particular, we are grateful to Murtaza Taj and Elisa Piccardo. Many thanks to Sushil Bhattacharjee for inspiring discussions and stimulating critical comments on drafts of the book. We thank Samuel Pachoud, Timothy Popkin, Nikola Spriljan, Toni Zgaljic and Huiyu Zhou for their contribution to the generation of test video sequences and for providing permission for the inclusion of sample frames in this publication. We are also grateful to colleagues who generated and made available relevant datasets for the evaluation of video trackers. These datasets are listed at: www.spevi.org. We thank the UK Engineering and Physical Sciences Research Council (EPSRC) and the European Commission that sponsored part of this work through the projects MOTINAS (EP/D033772/1) and APIDIS (ICT-216023). We owe particular thanks to Rama Chellappa for contributing the Foreword of this book. We are grateful to Nicky Skinner, Alex King and Clarissa Lim of John Wiley & Sons and Shalini Sharma of Aptara for their precious support in this project. Their active and timely cooperation is highly appreciated. Finally, we want to thank Silvia and Maria for their continuous encouragement and sustained support.

NOTATION

(w, h) target width and height

E_o single-target observation/measurement space

E_s single-target state space

E_I image space

I_k image at time index k

$M(k)$ number of targets at time index k

$N(k)$ number of measurements at time index k

X_k multi-target state

Z_k multi-target measurement

(u, v) target centroid

$\mathbf{X}_k$ set of active trajectories at time index k

$\mathbf{Z}_k$ set of measurements assigned to the active trajectories at time index k

$\mathbf{x}$ the collection of states (i.e. the time series) forming the target trajectory

$\mathcal{D}_{k|k}(x)$ Probability hypothesis density at time index k

$\mathcal{F}(E)$ collection of all finite subsets of the elements in E

$f_{k|k-1}$ state transition *pdf*

k time index (frame)

$p_{k-1|k-1}$ prior *pdf*

$p_{k|k-1}$ predicted *pdf*

$p_{k|k}$ posterior *pdf*

q_k importance sampling function

x_k state of a target at time index k

$y_{a:b}$ collection of elements (e.g, scalars, vectors) $\{y_a, y_{a+1}, \ldots, y_b\}$

z_k single-target measurement at time index k

g_k likelihood

ACRONYMS

AMF-PFR Adaptive multi-feature particle filter

AMI Augmented Multiparty Interaction

APIDIS Autonomous Production of Images Based on Distributed and Intelligent Sensing

CAVIAR Context Aware Vision using Image-based Active Recognition

CCD Charge-Coupled Device

CHIL Computers in the Human Interaction Loop

CIE Commission Internationale de l'Eclairage

CIF Common Intermediate Format (352×288 pixels)

CLEAR Classification of Events, Activities and Relationships

CNR Consiglio Nazionale delle Ricerche

CONDENSATION Conditional density propogation

CREDS Challenge of Real-time Event Detection Solutions

DoG Difference of Gaussians

EKF Extended Kalman filter

EM Expectation Maximisation

ETISEO Evaluation du Traitement et de l'Interpretation des Sequences vidEO

FISS Flinite Set Statistics

fps frames per second

GM-PHD Gaussian Mixture Probability Hypothesis Density

GMM Gaussian Mixture Model

HD High Definition

HY Hybrid-particle-filter-mean-shift tracker

i-Lids Imagery Library for Intelligent Detection Systems

KLT Kanade Lucas Tomasi

LoG Laplacian of Gaussian

MAP Maximum A Posteriori

MCMC Markov Chain Monte Carlo

MF-PF Multi-Feature Particle Filter

MF-PFR Multi-Feature Particle Filter with ad hoc Re-sampling

MHL Multiple-Hypothesis Localisation

MHT Multiple-Hypothesis Tracker

ML Maximum Likelihood

MODA Multiple Object Detection Accuracy

MODP Multiple Object Detection Precision

MOTA Multiple Object Tracking Accuracy

MOTP Multiple Object Tracking Precision

MS Mean Shift

MT Multiple-target tracker

PAL Phase Alternating Line

PCA Principal Components Analysis

PETS Performance Evaluation of Tracking and Surveillance

PF Particle Filter

PF-C Particle Filter, CONDENSATION implementation

PHD Probability Hypothesis Density

PHD-MT Multiple target tracker based on the PHD filter

PTZ Pan Tilt and Zoom

RFS Random Finite Set

RGB Red Green and Blue

SDV State-Dependent Variances

SECAM Sequentiel Couleur À Memoire

SHL Single-Hypothesis Localisation

SIFT Scale-Invariant Feature Transform

SPEVI Surveillance Performance EValuation Initiative

VACE Video Analysis and Content Extraction

ViPER Video Performance Evaluation Resource

1

WHAT IS VIDEO TRACKING?

1.1 INTRODUCTION

Capturing video is becoming increasingly easy. Machines that see and understand their environment already exist, and their development is accelerated
by advances both in micro-electronics and in video analysis algorithms. Now,
many opportunities have opened for the development of richer applications in
various areas such as video surveillance, content creation, personal communications, robotics and natural human–machine interaction.

One fundamental feature essential for machines to see, understand and
react to the environment is their capability to detect and track objects of
interest. The process of estimating over time the location of one or more
objects using a camera is referred to as *video tracking*. The rapid improvement
both in quality and resolution of imaging sensors, and the dramatic increase
in computational power in the past decade have favoured the creation of new
algorithms and applications using video tracking.

Video Tracking: Theory and Practice. Emilio Maggio and Andrea Cavallaro
© 2011 John Wiley & Sons, Ltd

Figure 1.1   Examples of targets for video tracking: (left) people, (right) faces.

The definition of object of interest[1] depends on the specific application at hand. For example, in a building surveillance application, targets may be people (Figure 1.1 (left)[2]), whereas in an interactive gaming application, targets may be the hands or the face of a person (Figure 1.1 (right)).

This chapter covers the fundamental steps for the design of a tracker and provides the mathematical formulation for the video tracking problem.

1.2   THE DESIGN OF A VIDEO TRACKER

Video cameras capture information about objects of interest in the form of sets of image pixels. By modelling the relationship between the appearance of the target and its corresponding pixel values, a video tracker estimates the location of the object over time.

The relationship between an object and its image projection is very complex and may depend on more factors than just the position of the object itself, thus making video tracking a difficult task. In this section, we first discuss the main challenges in video tracking and then we review the main components into which a video-tracking algorithm can be decomposed.

1.2.1   Challenges

The main challenges that have to be taken into account when designing and operating a tracker are related to the similarity of appearance between the target and other objects in the scene, and to appearance variations of the target itself.

---

[1] Note that the terms *target* and *object of interest* will be used interchangeably in this book.
[2] Unmarked image from the CAVIAR dataset (project IST-2001-37540).

Figure 1.2 Examples of clutter in video tracking. Objects in the background (red boxes) may share similar colour (left) or shape (right) properties with the target and therefore distract the tracker from the desired object of interest (green boxes). Left: image from the Birchfield head-tracking dataset. Right: Surveillance scenario from PETS-2001 dataset.

The appearance of other objects and of the background may be similar to the appearance of the target and therefore may interfere with its observation. In such a case, image features extracted from non-target image areas may be difficult to discriminate from the features that we expect the target to generate. This phenomenon is known as *clutter*. Figure 1.2 shows an example of colour ambiguity that can distract a tracker from the real target. This challenge can be dealt with by using multiple features weighted by their reliability (see Chapter 6).

In addition to the tracking challenge due to clutter, video tracking is made difficult by changes of the target appearance in the image plane that are due to one or more of the following factors:

- *Changes in pose.* A moving target varies its appearance when projected onto the image plane, for example when rotating (Figure 1.3(a)–(b)).

- *Ambient illumination.* The direction, intensity and colour of the ambient light influence the appearance of the target. Moreover, changes in global illumination are often a challenge in outdoor scenes. For example, ambient light changes when clouds obscure the sun. Also, the angles between the light direction and the normals to the object surface vary with the object pose, thus affecting how we see the object through the camera lens.

- *Noise.* The image acquisition process introduces into the image signal a certain degree of noise, which depends on the quality of the sensor. Observations of the target may be corrupted and therefore affect the performance of the tracker.

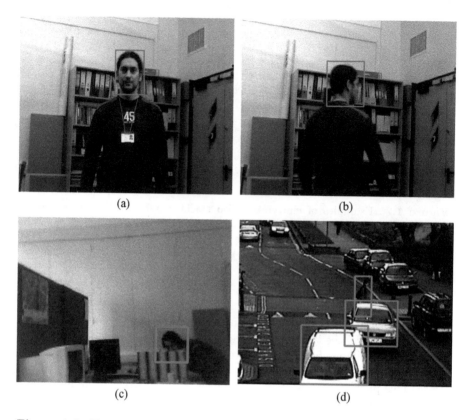

$\qquad$(a)$\qquad\qquad\qquad\qquad\qquad\qquad$(b)

$\qquad$(c)$\qquad\qquad\qquad\qquad\qquad\qquad$(d)

Figure 1.3   Examples of target appearance changes that make video tracking difficult. (a)–(b) A target (the head) changes its pose and therefore its appearance as seen by the camera. Bottom row: Two examples of target occlusions. (c) The view of the target is occluded by static objects in the scene. (d) The view of the target is occluded by another moving object in the scene; reproduced with permission of HOSDB.

- *Occlusions.* A target may fail to be observed when partially or totally occluded by other objects in the scene. Occlusions are usually due to:

  - a target moving behind a static object, such as a column, a wall, or a desk (Figure 1.3(c)), or

  - other moving objects obscuring the view of a target (Figure 1.3(d)).

  To address this challenge, different approaches can be applied that depend on the expected level of occlusion:

  - *Partial occlusions* that affect only a small portion of the target area can be dealt with by the target appearance model or by the target detection algorithm itself (see Section 4.3). The invariance properties of some global feature representation methods (e.g. the histogram) are appropriate to deal with occlusions. Also, the replacement of a global representation with multiple localised features that encode information for a small region of the target may increase the robustness of a video tracker.

– Information on the target appearance is not sufficient to cope with *total occlusions*. In this challenging scenario track continuity can be achieved via higher-level reasoning or through multi-hypothesis methods that keep propagating the tracking hypotheses over time (see Section 5.3). Information about typical motion behaviours and pre-existing occlusion patterns can also be used to propagate the target trajectory in the absence of valid measurements. When the target reappears from the occlusion, the propagation of multiple tracking hypotheses and appearance modelling can provide the necessary cues to reinitialise a track.

A summary of the main challenges in video tracking is presented in Figure 1.4.

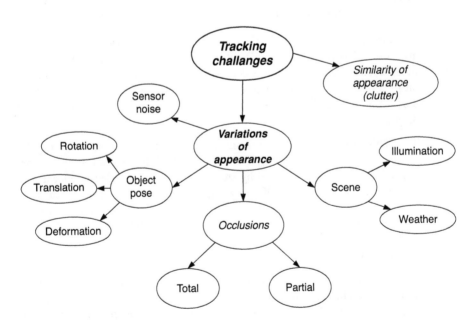

Figure 1.4 The main challenges in video tracking are due to temporal variations of the target appearance and to appearance similarity with other objects in the scene.

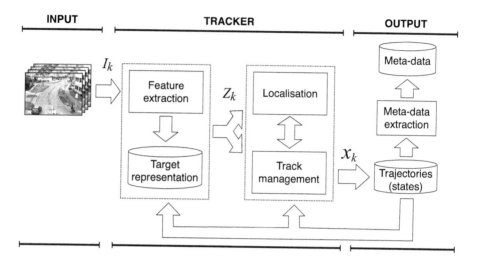

Figure 1.5 The video-tracking pipeline. The flow chart shows the main logical components of a tracking algorithm.

1.2.2 Main components

In order to address the challenges discussed in the previous section, we identify five main logical components of a video tracker (Figure 1.5):

1. The definition of a method to *extract* relevant information from an image area occupied by a target (Chapter 3). This method can be based on motion classification, change detection, object classification or simply on extracting low-level features such as colour or gradient (see Section 3.3), or mid-level features such as edges or interest points (see Section 3.4).

2. The definition of a representation for encoding the appearance and the shape of a target (the *state*). This representation defines the characteristics of the target to be used by the tracker (Chapter 4). In general, the representation is a trade-off between accuracy of the description (descriptiveness) and invariance: it should be descriptive enough to cope with clutter and to discriminate false targets, while allowing a certain degree of flexibility to cope with changes of target scale, pose, illumination and partial occlusions (see Sections 4.2 and 4.3).

3. The definition of a method to *propagate* the state of the target over time. This step recursively uses information from the feature extraction step or from the already available state estimates to form the trajectory (see Chapter 5). This task links different instances of the same object over time and has to compensate for occlusions, clutter, and local and global illumination changes.

4. The definition of a strategy to manage targets appearing and disappearing from the imaged scene. This step, also referred to as *track management*, initialises the track for an incoming object of interest and terminates the trajectory associated with a disappeared target (see Chapter 7). When a new target appears in the scene (*target birth*), the tracker must initialise a new trajectory. A target birth usually happens:

- at the image boundaries (at the edge of the field of view of the camera),

- at specific entry areas (e.g. doors),

- in the far-field of the camera (when the size of the projection onto the image plane increases and the target becomes visible), or

- when a target spawns from another target (e.g. a driver parking a car and then stepping out).

Similarly, a trajectory must be terminated (*target death*) when the target:

- leaves the field of view of the camera, or

- disappears at a distance or inside another object (e.g. a building).

In addition to the above, it is desirable to terminate a trajectory when the tracking performance is expected to degrade under a predefined level, thus generating a *track loss* condition (see Section 9.5.1).

5. The extraction of *meta-data* from the state in a compact and unambiguous form to be used by the specific application, such as video annotation, scene understanding and behaviour recognition. These applications will be described in Chapter 2.

In the next sections we will discuss in detail the first four components and specific solutions used in popular video trackers.

1.3 PROBLEM FORMULATION

This section introduces a formal general definition of the video-tracking problem that will be used throughout the book. We first formulate the single-target tracking problem and then extend the definition to multiple simultaneous target tracking.

1.3.1 Single-target tracking

Let $\mathbf{I} = \{I_k : k \in \mathbb{N}\}$ represent the frames of a video sequence, with $I_k \in E_I$ being the frame (image plane) at time k, defined in E_I, the space of all possible images.

Tracking a single target using monocular video can be formulated as the estimation of a time series

$$\mathbf{x} = \{x_k : k \in \mathbb{N}\} \tag{1.1}$$

over the set of discrete time instants indexed by k, based on the information in $\mathbf{I}$. The vectors $x_k \in E_s$ are the *states* of the target and E_s is the state space. The time series $\mathbf{x}$ is also known as the *trajectory* of the target in E_s. The information encoded in the state x_k depends on the application.

I_k may be mapped onto a feature (or observation) space E_o that highlights information relevant to the tracking problem. The observation generated by a target is encoded in $z_k \in E_o$. In general, E_o has a lower dimensionality than that of original image space, E_I (Figure 1.6).

The operations that are necessary to transform the image space E_I to the observation space E_o are referred to as *feature extraction* (see Chapter 3). Video trackers propagate the information in the state x_k over time using the extracted features. A localisation strategy defines how to use the image features to produce an estimate of the target state x_k (see Chapter 5).

We can group the information contained in x_k into three classes:

1. Information on the target *location* and *shape*. The positional and shape information depends on the type of object we want to track and on the amount (and quality) of the information we can extract from the images. We will discuss shape approximations for tracking in Section 4.2.

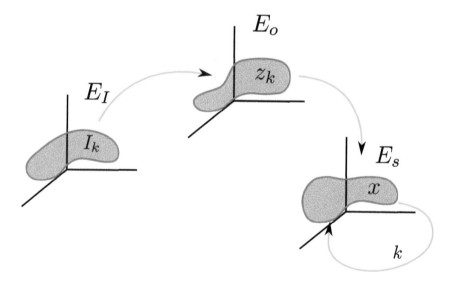

Figure 1.6 The flow of information between vector spaces in video tracking. The information extracted from the images is used to recursively estimate the state of the target (Key. E_I: the space of all possible images; E_o: feature or observation space; E_s: state space; k: time index).

2. Information on the target *appearance*. Encoding appearance information in the state helps in modelling appearance variations over time. We will cover target appearance representations in Section 4.3.

3. Information on the *temporal variation* of shape or appearance. The parameters of this third class are usually first- or higher-order derivatives of the other parameters, and are optional. The description of shape and appearance variations in the state will be discussed in Section 4.3.3.

Note that some elements of the state x_k may not be part of the final output required by the specific application. This extra information is used as it may be beneficial to the performance of the tracker itself. For example, tracking appearance variations through a set of state parameters may help in coping with out-of-plane rotations. Nevertheless, as adding parameters to the state increases the complexity of the estimator, it is usually advisable to keep the dimensionality of x_k as low as possible.

Figure 1.7 shows examples of states describing location and an approximation of the shape of a target. When the goal is tracking an object on the

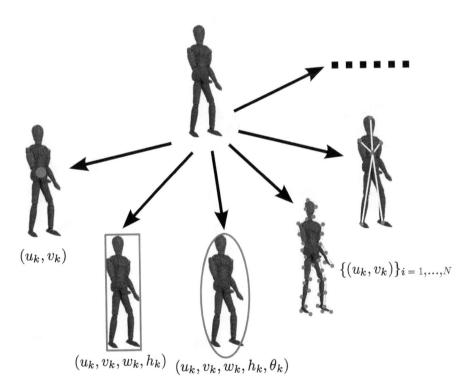

(u_k, v_k)

$\{(u_k, v_k)\}_{i=1,\dots,N}$

(u_k, v_k, w_k, h_k) $(u_k, v_k, w_k, h_k, \theta_k)$

Figure 1.7   Example of state definitions for different video-tracking tasks.

image plane, the minimal form of x_k will represent the position of a point in I_k, described by its vertical and horizontal coordinates, that is

$$x_k = (u_k, v_k). \tag{1.2}$$

Similarly, one can bound the target area with a rectangle or ellipse, defining the state x_k as

$$x_k = (u_k, v_k, h_k, w_k, \theta_k), \tag{1.3}$$

where $y_k = (u_k, v_k)$ defines the centre, h_k the height, w_k the width and (optionally) θ_k the clockwise rotation. More complex representations such as chains of points on a contour can be used. Chapter 4 will provide a comprehensive overview of shape and appearance representations.

1.3.2   Multi-target tracking

When the goal of video tracking is to simultaneously track a time-varying number of targets, we can formulate the multi-target state X_k as the vector containing the concatenation of the parameters of the single-target states. If $M(k)$ is the number of targets in the scene at time k, and $\mathcal{F}(E)$ is the collection of all the finite subsets of E, then the multi-target state, X_k, is the set

$$X_k = \{x_{k,1}, \ldots x_{k,M(k)}\} \in \mathcal{F}(E_s). \tag{1.4}$$

Some multi-target tracking algorithms solve the tracking problem as the association of multiple observations generated by an object detector over time (see Section 3.5). In this case the single-target observation vector $z_k \in E_o$ is composed of parameters defining a single detection. Typically these parameters represent the position and size of a bounding box. Similarly to Eq. (1.4) we can extend the definition of observation to multiple targets by defining the multi-target observation (or measurement) Z_k as the finite collection of the single target observations, that is

$$Z_k = \{z_{k,1}, \ldots z_{k,N(k)}\} \in \mathcal{F}(E_o), \tag{1.5}$$

formed by the $N(k)$ observations. For simplicity, let us define the set of active trajectories at frame k as

$$\mathbf{X}_k = \{\mathbf{x}_{k,1}, \ldots \mathbf{x}_{k,M(k)}\}, \tag{1.6}$$

and

$$\mathbf{Z}_k = \{\mathbf{z}_{k,1}, \ldots \mathbf{z}_{k,M(k)}\}, \tag{1.7}$$

where each $\mathbf{z}_{k,i}$ is the set of observations assigned to the trajectory i up to time step k. Multi-target tracking algorithms will be covered extensively in Chapter 7.

1.3.3 Definitions

In order to evaluate the quality of the state estimation, we will quantify the accuracy and the precision of tracking results:

- The *accuracy* measures the closeness of the estimates to the real trajectory of the target.

- The *precision* measures the amount of variation in the trajectory estimates across repeated estimations that are obtained under the same conditions. In other words, the precision measures the degree of repeatability of a video-tracking result.

We will discuss the formal evaluation of the state estimation results in Chapter 9. The *state* definitions that we will use throughout the book are listed below. Similar definitions and notations are valid for representing the *measurements* extracted from the images:

- x: a vector in the state space

- x_k: state of a target at time index k

- $x_{a:b}$: the collection of states between time indexes a and b

- $\mathbf{x}$: the collection of states (i.e. the time series) forming the trajectory

- $\mathbf{x}_k$: the collection of all the states forming the trajectory of a target up to time index k

- $x_{k,j}$: the state of the jth target at time index k

- $x_{a:b,j}$: the collection of states of the jth target between time indexes a and b

- $\mathbf{x}_{k,j}$: collection of all the states forming the trajectory of the jth target up to time index k

- X: a set of vectors in the single-target state space

- $X_k = \{x_{k,1}, x_{k,2}, \ldots, x_{k,M(k)}\}$: the multi-target state at time k

- $X_{a:b} = \{x_{a:b,1}, x_{a:b,2}, \ldots\}$: the set of trajectory states between time indexes a and b

- $\mathbf{X}_k = \{\mathbf{x}_{k,1}, \mathbf{x}_{k,2}, \ldots\}$: the set of trajectory states up to time k.

Figure 1.8 shows a pictorial representation of the symbols defined above and used for a multi-target state.

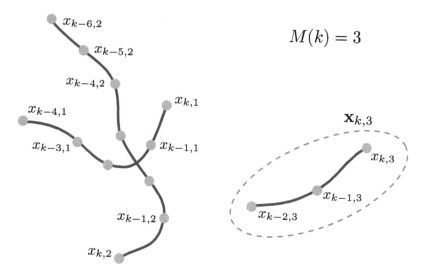

$$M(k) = 3$$

Figure 1.8 Pictorial representation of three trajectories: definitions of the symbols used for a multi-target state.

1.4 INTERACTIVE VERSUS AUTOMATED TRACKING

Based on the amount of interaction with the users to identify the object(s) of interest, video-tracking algorithms can be grouped into three classes, namely manual, interactive (or supervised) and automated (or unsupervised). These three classes are discussed below:

- Video tracking may be performed directly by the user. *Manual* tracking is used when a high accuracy, for example in the definition of the boundaries of the targets, is necessary. One example of application that requires manual tracking is in film production, when the contour of a character is selected and its evolution over time is defined by an operator, on each frame. Although this procedure allows for a good definition of the target position (or boundaries), it is very time consuming and cannot be extensively used for large volumes of visual data.

- *Automated* tracking uses a priori information about targets that is encoded in an algorithm. For example, an object detector can be used to initialise the tracker and/or to support the state estimation over time. Examples of automated tracking methods are those based on face detection and moving-object segmentation (see Section 3.5). Fully automated video-tracking techniques for real-world applications are still in their infancy, because translating the properties defining a generic target into algorithmic criteria is a difficult task in non-constrained scenes.

- *Interactive* (semi-automated) strategies are used as a trade-off between a fully automated tracker and a manual tracker. The principle at the

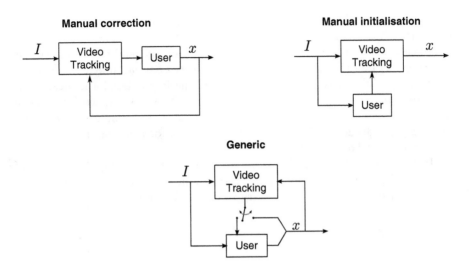

Figure 1.9 Automated and interactive tracking. Block diagrams representing different configurations enabling user interaction in video tracking.

basis of semi-automated techniques is the interaction of the user during some stages of the tracking process, where the information about the target area is provided directly by the user. Interactive tracking is used in *tag-and-track* applications (e.g. video editing and surveillance) when an operator manually initialises (selects) a target of interest that is then followed by the tracking algorithm. After the user provides the initial definition of the target, the tracker follows its temporal evolution in the subsequent frames thus propagating the initialisation information. This second phase can be either unsupervised, or the operator can still verify the quality of the tracking results and correct them if needed. This allows, for example, video editors or annotators to select the appearance of an object in a single frame and then to propagate these changes over time, thus saving the time-consuming task of manually editing each frame. Another example of an application is camera control for surveillance, when a person is selected by an operator and then automatically followed with a pan tilt and zoom (PTZ) camera.

Schematic representations of different types of interactive procedure are shown in Figure 1.9.

1.5 SUMMARY

In this chapter we introduced the concept of video tracking and discussed the major steps in the design of a tracker. We defined a tracking taxonomy based on five building blocks that can be commonly identified in video-tracking algorithms. In this context, we highlighted the factors that make video tracking a difficult task. To cope with these factors, a tracker should have an invariant representation of the target or else adapt the representation over time.

Next, we introduced the definition and a generic problem formulation that is based on the concept of the state of a target. Moreover, we discussed a classification of tracking algorithms that is based on the amount of interaction with a user. Other classifications are possible as, for example, marker-based and markerless tracking or causal and non-causal tracking.

In the remainder of the book we will discuss the fundamental aspects of video-tracking algorithm design and the various implementation choices and trade-off decisions that are necessary to define accurate video trackers.

2

APPLICATIONS

2.1 INTRODUCTION

Tracking objects of interest in video is at the foundation of many applications, ranging from video production to remote surveillance, and from robotics to interactive immersive games. Video trackers are used to *improve our understanding* of large video datasets from medical and security applications; to *increase productivity* by reducing the amount of manual labour that is necessary to complete a task and to enable *natural interaction* with machines.

In this chapter we offer an overview of current and upcoming applications that use video tracking. Although the boundaries between these applications are somehow blurred, they can be grouped in six main areas:

- Media production and augmented reality

- Medical applications and biological research

- Surveillance and business intelligence

- Robotics and unmanned vehicles

- Tele-collaboration and interactive gaming

- Art installations and performances.

Specific examples of these applications will be covered in the following sections.

2.2   MEDIA PRODUCTION AND AUGMENTED REALITY

Video tracking is an important element in post-production and motion capture for the movie and broadcast industries. *Match moving* is the augmentation of original shots with additional computer graphics elements and special effects, which are rendered in the movie. In order to consistently add these new elements to subsequent frames, the rendering procedure requires the knowledge of 3D information on the scene. This information can be estimated by a *camera tracker*, which computes over time the camera position, orientation and focal length. The 3D estimate is derived from the analysis of a large set of 2D trajectories of salient image features that the video tracking algorithm identifies in the frames [1, 2]. An example of tracking patches and points is shown in Figure 2.1, where low-level 2D trajectories are used to estimate higher-level 3D information. Figure 2.2 shows two match-moving examples where smoke special effects and additional objects (a boat and a building) are added to the real original scenes. A related application is *virtual product placement* that includes a specific product to be advertised in a video or wraps a logo or a specific texture around an existing real object captured in the scene.

Figure 2.1   Example of a camera tracker that uses the information obtained by tracking image patches. Reproduced with permission of the Oxford Metrics Group.

Figure 2.2 Match-moving examples for special effects and object placement in a dynamic scene. Top: smoke and other objects are added to a naval scene. Bottom: the rendering of a new building is added to an aerial view. Reproduced with permission of the Oxford Metrics Group.

Another technology based on video tracking and used by media production houses is *motion capture*. Motion capture systems are used to animate virtual characters from the tracked motion of real actors. Although markerless motion capture is receiving increasing attention, most motion-capture systems track a set of markers attached to an actor's body and limbs to estimate their poses (Figure 2.3). Specialised motion-capture systems recover the movements of real actors in 3D from the tracked markers. Then the motion of the makers is mapped onto characters generated by computer graphics.

Video tracking is also used for the analysis and the enhancement of sport events. As shown in the example of Figure 2.4, a tracking algorithm can estimate the position of players in the field in order to gather statistics about a game (e.g. a football match). Statistics and enhanced visualisations aid the commentators, coaches and supporters in highlighting team tactics and player performance.

2.3 MEDICAL APPLICATIONS AND BIOLOGICAL RESEARCH

The motion-capture tools described in the previous section are also used for the analysis of human motion to improve the performance of athletes

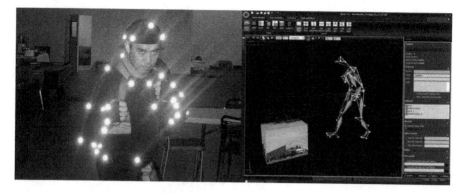

Figure 2.3 Examples of motion capture using a marker-based system. Left: retro-reflective markers to be tracked by the system. Right: visualisation of the motion of a subject. Reproduced with permission of the Oxford Metrics Group.

(Figure 2.5(a)–(b)) and for the analysis of the gait of a patient [3] to assess the condition of the joints and bones (Figure 2.5(c)). In general, video tracking has been increasingly used by medical systems to aid the diagnosis and to speed up the operator's task. For example, automated algorithms track the ventricular motion in ultrasound images [4–6]. Moreover, video tracking can estimate the position of particular soft tissues [7] or of instruments such as needles [8, 9] and bronchoscopes [10] during surgery.

In *biological research*, tracking the motion of non-human organisms allows one to analyse and to understand the effects of specific drugs or the effects of ageing [11–15]. Figure 2.6 shows two application examples where video

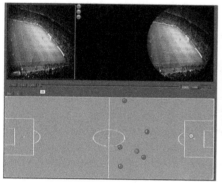

Figure 2.4 Video tracking applied to media production and enhanced visualisation of sport events. Animations of the real scene can be generated from different view-points based on tracking data. Moreover, statistics regarding player positions are automatically gathered and may be presented as overlay or animation. Reproduced with permission of Mediapro.

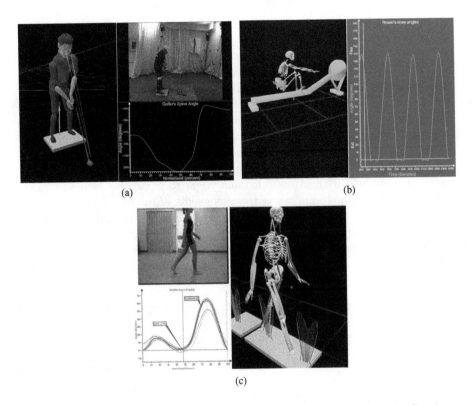

Figure 2.5 Example of video tracking for medical and sport analysis applications. Motion capture is used to analyse the performance of a golfer (a), of a rower (b) and to analyse the gait of a patient (c). Reproduced with permission of the Oxford Metrics Group.

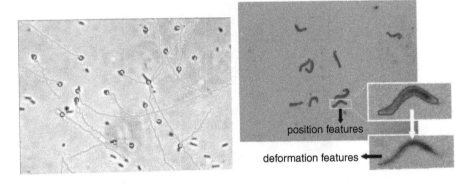

Figure 2.6 Examples of video tracking for medical research. Automated tracking of the position of *Escherichia coli* bacteria (left) and of *Caenorhabditis elegans* worms (right). Left: reproduced from [14]; right: courtesy of Gavriil Tsechpenakis, IUPUI.

tracking estimates the location of *Escherichia coli* bacteria and *Caenorhabditis elegans* worms.

2.4 SURVEILLANCE AND BUSINESS INTELLIGENCE

Video tracking is a desirable tool used in automated *video surveillance* for security, assisted living and business intelligence applications. In surveillance systems, tracking can be used either as a forensic tool or as a processing stage prior to algorithms that classify behaviours [16]. Moreover, video-tracking software combined with other video analytical tools can be used to redirect the attention of human operators towards events of interest.

Smart surveillance systems can be deployed in a variety of different indoor and outdoor environments such as roads, airports, ports, railway stations, public and private buildings (e.g. schools, banks and casinos). Examples of video surveillance systems (Figure 2.7) are the IBM Smart Surveillance System

(a) (b)

(c)

Figure 2.7 Examples of object tracking in surveillance applications. (a)–(b): General Electric intelligent video platform; (c): ObjectVideo surveillance platform. The images are reproduced with permission of General Electric Company (a,b) and ObjectVideo (c).

Figure 2.8 Examples of video tracking for intelligent retail applications. Screen shots from IntelliVid software (American Dynamics).

(S3) [17, 18], the General Electrics VisioWave Intelligent Video Platform [19] and Object Video VEW [20, 21].

Video tracking may also serve as an observation and measurement tool in retail environments (e.g. *retail intelligence*), such as supermarkets, where the position of customers is tracked over time [22] (Figure 2.8). Trajectory data combined with information from the point of sales (till) is used to build behavioural models describing where customers spend their time in the shop, how they interact with products dpending on their location, and what items they buy. By analysing this information, the marketing team can improve the product placement in the retail space. Moreover, gaze tracking in front of billboards can be used to automatically select the type of advertisment to show or to dynamically change its content based on the attention or the estimated marketing profile of a person, based for example on the estimated gender and age.

2.5 ROBOTICS AND UNMANNED VEHICLES

Another application area that extensively uses video-tracking algorithms is robotics. Robotic technology includes the development of humanoid robots, automated PTZ cameras and unmanned aerial vehicles (UAVs). Intelligent vision via one or more cameras mounted on the robots provide information that is used to interact with or navigate in the environment. Also environment exploration and mapping [23], as well as human–robot interaction via gesture recognition rely on video tracking [24].

The problem of estimating the global motion of robots and unmanned vehicles is related to the camera-tracking problem discussed in Section 2.2. While tracking algorithms for media production can be applied off-line, video trackers for robotics need to simultaneously localise in realtime the position of the robot (i.e. of the camera) and to generate a map of the environment. 3D localisation information is generated by tracking the position of prominent image features such as corners and edges [25, 26], as shown in Figure 2.1.

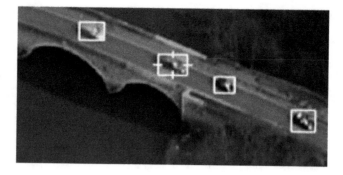

Figure 2.9 Example of object tracking from an Unmanned Aerial Vehicle. Reproduced with permission of the Oxford Metrics Group.

Information on the 3D position is also used to generate a 3D mesh approximating the structure of surrounding objects and the environment. In particular, UAVs make extensive use of video tracking to find the position of specific objects on the ground (Figure 2.9) as well as to enable automated landing.

2.6   TELE-COLLABORATION AND INTERACTIVE GAMING

Standard webcams are already shipped with tracking software that localises and follows the face of a user for *on-desk video conferencing*. Moreover, video-based gaze tracking is used to simulate eye contact among attendees of a meeting to improve the effectiveness of interaction in video-conferencing [27]. Video tracking technology for lecture rooms is available that uses a set of PTZ cameras to follow the position of the lecturer [27–30]. The PTZ cameras exploit the trajectory information in real-time to guide the pan, tilt and zoom parameters of the camera. To improve tracking accuracy, information from an array of microphones may also be fused with the information from the camera [31].

Video tracking is also changing the way we send *control* to machines. This natural interaction modality is being used in interactive games. For example, the action of pressing a button on the controller is replaced by a set of more intuitive gestures performed by the user in front of the camera [32] (Figure 2.10). Likewise, in *pervasive games*, where the experience extends to the physical world, vision-based tracking refines positional data from the Global Positioning System (GPS) [33].

2.7   ART INSTALLATIONS AND PERFORMANCES

Video tracking is increasingly being used in art installations and performances where interaction is enabled by the use of video cameras and often by projection systems. The interactivity can be used to enhance the narrative of a piece or to create unexpected actions or reactions of the environment.

Figure 2.10 Examples of gesture interfaces for interactive gaming. The video camera uses tracking technology to capture the gestures that are converted into gaming actions. Reproduced with permission of GestureTek.

For example, tracking technology enables interaction between museum goers and visual installations (Figure 2.11, left). Also, someone in a group can be selectively detected and then tracked over time while a light or an 'animated' shadow is projected next to the selected person (Figure 2.11, right).

Interactive art based on video tracking can also enable novel forms of communication between distant people. For example the relative position of a tracked object and a human body may drive a set of lighting effects [34].

2.8 SUMMARY

In the past few decades an increasing effort driven by the creative, robotics, defence and surveillance industries has been devoted to the development of

Figure 2.11 Examples of video tracking applied to interactive art installations. A person interacts with objects visualised on a large display (left, reproduced with permission of Alterface). An animated shadow (right) appears next to a person being tracked.

specialised video trackers, which are important components in a wide range of applications. This chapter discussed a series of representative application examples in the field of video editing, interpretative surveillance, motion capture and human–machine interactive systems. As the reliability of the algorithms progresses and the type and variety of applications is ever increasing, an updated list of applications and their corresponding description is available at www.videotracking.org.

REFERENCES

1. B. Triggs, P.F. McLauchlan, R.I. Hartley and A.W. Fitzgibbon. Bundle adjustment - a modern synthesis. In *Proceedings of the International Conference on Computer Vision*, London, UK, 1999, 298–372.

2. J. Shi and C. Tomasi. Good features to track. In *Proceedings of the IEEE Conference on Computer Vision and Pattern Recognition*, Seattle, USA, 1994, 593–600.

3. I. Charlton, P. Tate, P. Smyth, and L. Roren. Repeatability of an optimised lower body model. *Gait & Posture*, 20(2):213–221, 2004.

4. E. Bardinet, L.D. Cohen and N. Ayache. Tracking and motion analysis of the left ventricle with deformable superquadrics. *Medical Image Analysis*, 1:129–149, 1996.

5. Y. Notomi, P. Lysyansky, R. Setser, T. Shiota, Z. Popovic, M. Martin-Miklovic, J. Weaver, S. Oryszak, N. Greenberg and R. White. Measurement of ventricular torsion by two-dimensional ultrasound speckle tracking imaging. *Journal of the American College of Cardiology*, 45(12):2034–2041, 2005.

6. Toshiba. Artida untrasound scanner. http://medical.toshiba.com/Products/US/Artida/. Last visited: 21 April 2009.

7. P. Mountney and G.-Z. Yang. Soft tissue tracking for minimally invasive surgery: Learning local deformation online. In *Proceedings of the International Conference on Medical Image Computing and Computer-Assisted Intervention*, Berlin, Germany, 2008, 364572.

8. D.A. Leung, J.F. Debatin, S. Wildermuth, N. Heske, C.L. Dumoulin, R.D. Darrow, M. Hauser, C.P. Davis, and GK Von Schulthess. Real-time biplanar needle tracking for interventional mr imaging procedures. *Radiology*, 197:485–488, 1995.

9. E. Kochavi, D. Goldsher and H. Azhari. Method for rapid MRI needle tracking. *Magnetic Resonance in Medicine*, 51(5):1083–1087, 2004.

10. K. Mori, D. Deguchi, K. Akiyama, T. Kitasaka, C.R. Maurer Jr., Y. Suenaga, H. Takabatake, M. Mori and H. Natori. Hybrid bronchoscope tracking using a magnetic tracking sensor and image registration. In *Proceedings of the International Conference on Medical Image Computing and Computer-Assisted Intervention*, Palm Springs CA, 2005, 543–550.

11. W. Geng, P. Cosman, C.C. Berry, Z. Feng, and W.R. Schafer. Automatic tracking, feature extraction and classification of *C elegans* phenotypes. *IEEE Transactions on Biomedical Engineering*, 51(10):1811–1820, 2004.

12. C. Restif and D. Metaxas. Tracking the swimming motions of *C. elegans* worms with applications in aging studies. In *Proceedings of the International Conference*

on *Medical Image Computing and Computer-Assisted Intervention*, New York, 2008, 35–42.

13. G. Tsechpenakis, L. Bianchi, D. Metaxas, and M. Driscoll. A novel computational approach for simultaneous tracking and feature extraction of *C. elegans* populations in fluid environments. *IEEE Transactions on Biomedical Engineering*, 55:1539–1549, 2008.

14. J. Xie, S. Khan and M. Shah. Automatic tracking of *Escherichia coli* bacteria. In *Proceedings of the International Conference on Medical Image Computing and Computer-Assisted Intervention*, Berlin, Germany, 2008, 824–832.

15. A. Veeraraghavan, R. Chellappa and M. Srinivasan. Shape and behavior encoded tracking of bee dances, *IEEE Transactions on Pattern Analysis and Machine Intelligence*, 30(1):463–476, 2008.

16. N. Anjum and A. Cavallaro. Multi-feature object trajectory clustering for video analysis. *IEEE Transactions on Circuits and Systems for Video Technology*, 18(11): 2008.

17. C.-F. Shu, A. Hampapur, M. Lu, L. Brown, J. Connell, A. Senior and Y. Tian. IBM Smart Surveillance System (S3): a open and extensible framework for event based surveillance. In *Proceedings of the IEEE Conference on Advanced Video and Signal Based Surveillance*, Como, Italy, 2005, 318–323.

18. IBM. IBM Smart Surveillance System (S3). http://www.research.ibm.com/peoplevision/2Dtrack.html. Last visited: 2 May 2009.

19. GE. General Electrics VisioWave Intelligent Video Platform. www.gesecurity.com. Last visited: 2 May 2009.

20. A.J. Lipton, J.I. Clark, P. Brewer, P.L. Venetianer and A.J. Chosak. Objectvideo forensics: activity-based video indexing and retrieval for physical security applications. *IEE Intelligent Distributed Surveilliance Systems*, 1:56–60, February 2004.

21. ObjectVideo. Object Video VEW. http://www.objectvideo.com/products/vew/. Last visited: 2 May 2009.

22. IntelliVid. IntelliVid retail Video Investigator[TM]. http://www.intellivid.com. Last visited: 2 May 2009.

23. A.J. Davison and N.D. Molton. Monoslam: Real-time single camera slam. *IEEE Transactions on Pattern Analysis and Machine Intelligence*, 29(6):1052–1067, 2007.

24. Y. Wu and T.S. Huang. Vision-based gesture recognition: A review. In *Proceedings of the International Gesture Workshop on Gesture-Based Communication in Human-Computer Interaction*, London, UK, 1999, 103–115.

25. A.J. Davison, I.D. Reid, N.D. Molton and O. Stasse. MonoSLAM: Real-time single camera SLAM. *IEEE Transactions on Pattern Analysis and Machine Intelligence*, 26(6):1052–1067, 2007.

26. A.W. Fitzgibbon and A. Zisserman. Automatic camera recovery for closed or open image sequences. In *Proceedings of the European Conference on Computer Vision*, London, UK, 1998, 311–326.

27. J. Gemmell, K. Toyama, C.L. Zitnick, T. Kang and S. Seitz. Gaze awareness for video-conferencing: a software approach. *IEEE Multimedia*, 7(4):26–35, 2000.

28. K.S. Huang and M.M. Trivedi. Video arrays for real-time tracking of person, head, and face in an intelligent room. *Machine Vision Applications*, 14(2):103–111, 2003.

29. CHIL. Computers in the Human Interaction Loop website. http://www.chil.server.de. Last visited: 18 March 2010.

30. C. Wang and M.S. Brandstein. Multi-source face tracking with audio and visual data. In *Proceedings of the IEEE Workshop on Multimedia Signal Processing*, Copenhagen, Denmark, 1999, 169–174.

31. W.T. Freeman, D.B. Anderson, P.A. Beardsley, C.N. Dodge, M. Roth, C.D. Weissman, W.S. Yerazunis, H. Kage, K. Kyuma, Y. Miyake and K. Tanaka. Computer vision for interactive computer graphics. *IEEE Computer Graphics and Applications*, 18(3):42–53, 1998.

32. I. Lindt and W. Broll. Netattack – first steps towards pervasive gaming. *ERCIM News*, 57, 2004.

33. T. Hayashi, S. Agamanolis and M. Karau. Mutsugoto: a body-drawing communicator for distant partners. In proceedings of *SIGGRAPH 2008: International Conference on Computer Graphics and Interactive Techniques*, Los Angeles, 2008, 1–1.

3

FEATURE EXTRACTION

3.1 INTRODUCTION

The performance of a video tracker depends on the quality of the information we can extract from the images. To understand how to better exploit image information, in this chapter we briefly introduce the image formation process and discuss methods to extract features that are significant for the disambiguation of the objects of interest:

- from the background scene and

- from other objects.

Feature extraction is the first step in the tracking pipeline and allows us to highlight information of interest from the images to represent a target. The features to be extracted can be grouped into three main classes, namely:

- Low-level (e.g. colour, gradient, motion)

- Mid-level (e.g. edges, corners, regions)

- High-level (objects).

Video Tracking: Theory and Practice. Emilio Maggio and Andrea Cavallaro
© 2011 John Wiley & Sons, Ltd

3.2 FROM LIGHT TO USEFUL INFORMATION

Video tracking exploits the information captured by an imaging sensor and generates a compact description of the localisation of objects of interest to be used for the application at hand. In this section we briefly discuss the image formation process that generates the input signal for a video tracker.

3.2.1 Measuring light

An imaging sensor is composed of an array of photosensitive elements, the pixels. When the light hits a photosensitive element, due to the photoelectric effect, the state of the element changes. This change is proportional to the number of photons (i.e. the intensity of the light) hitting the element.

The photosensitive element works as a capacitor. We can obtain a quantitative estimate of the light intensity by measuring the voltage at the output pins of the photo element. For example, in grey-level sensors, zero voltage (i.e. no photons) means black, while the maximum voltage is produced as a consequence of many photons hitting the sensor (i.e. white).

Advancements in silicon technology allow for the integration of millions of pixels on a single microchip where the pixels form a two-dimensional array (Figure 3.1). The number of pixels of a sensor is also known as its *resolution*. Due to the small size of the sensor (usually from 1 to 10 millimetres), a lens is needed to concentrate the light on the array so that each pixel encodes the information from different directions within the view angle of the lens.

3.2.1.1 The colour sensor   Colour data acquisition follows a similar procedure to that described above. However, differences exist between devices that

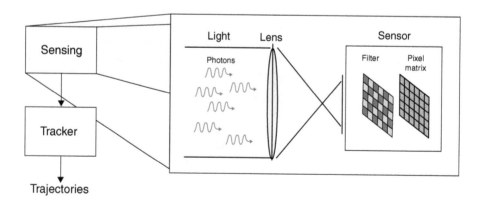

Figure 3.1 Image sensing for video tracking. The light reflected by the objects within the camera's field of view is converted into digital information that is used as input for the tracker.

use one or multiple sensors:

- In colour devices that use a *single sensor*, different filters are positioned in front of each pixel, as shown in Figure 3.1, to form an array. The filters let through the light within a range of frequencies. Usually the array is composed of three different filters that are sensitive to three different frequency bands. After analogue-to-digital conversion, the output of neighbouring physical pixels is interpolated to generate a virtual representation where each colour pixel is a triplet corresponding to the red, green and blue colour components. Although single-sensor devices are cheaper than multi-sensor devices, more than 60% of the information carried by the light is lost during the filtering process, thus resulting in suboptimal performance in terms of picture quality. Moreover, the effective resolution is smaller than the number of physical pixels and interpolation in each colour plane is necessary to reach the physical resolution of the sensor.

- Colour devices that use *multiple sensors* better exploit the photons passing through the lens. The filter array used in single-sensor devices is here substituted by a prism that splits the photons according to their frequency (i.e. their associated colour), thus forming multiple light beams. Each beam is directed towards a separate sensor. The three charge coupled device camera, the most popular multi-sensor device, uses a trichroic prism (i.e. three beams) paired with three sensors. Unlike single-sensor devices, in multi-sensor devices the interpolation of colour information is not necessary and therefore the effective resolution is equal to the physical resolution of each sensor.

A large variety of colour sensors exist that differ in terms of filter colour bands, density of pixels per band and pixel arrangement. Also, sensors may operate on frequency bands beyond the visible spectrum (e.g. infra red cameras) or across multiple frequency bands of the electromagnetic spectrum (i.e. multi-spectral and hyper-spectral cameras).

As a complete treatment of sensor technology is outside the scope of this book, interested readers can refer to the book by Holts and Lomheim [1] for an in-depth coverage.

3.2.1.2 *From images to videos*

Digital video cameras activate the electronic shutter multiple times per second, thus generating a sequence of images depicting the scene at different time instants. The number of activations per seconds defines the *frame rate* of the camera. The frame rate varies from a few frames per second (fps) to hundreds or thousands fps for special purpose cameras [2].

From a video-tracking perspective, the camera frame rate is a very important parameter: higher frame rates facilitate tracking as, at equal object speed, the magnitude of the motion between two frames is smaller. Therefore, simpler but faster algorithms that can run at high frame rates may perform better than more elaborate but slower methods. Also, fast shutter speed used with high-frame-rate cameras reduces *motion blur*, thus improving the sharpness of the images with fast-moving objects.

3.2.1.3 *Representing the video*   If we disregard the quantum effects due to the discrete nature of the light, the voltage is still an analogue signal spanning a range between zero and its maximum. To obtain a digital signal, an analogue-to-digital converter maps the voltage onto a finite set of binary numbers. For example, in many commercial cameras, the voltage is converted into an 8-bit unsigned integer in the range 0–255. The collection of bits representing an image is often referred to as a *raw* or *uncompressed image*. Similarly, a *raw video* is the concatenation of the raw images representing subsequent frames.

The number of bits necessary to represent a raw image depends on two factors: the image resolution and the colour depth, measured in bits per pixel. The colour depth depends on the number of quantisation levels. In addition to these two factors, in a raw video the number of bits per second also depends on the frame rate.

The bandwidth and storage space required by raw video quickly grow with increasing sensor resolution and quantisation levels. Consequently, the management of raw videos is impractical in many applications and video compression techniques are often used. Video coders aim to reduce the number of bits that are necessary to represent a raw video by exploiting spatial and temporal redundancies between neighbouring pixels in the same frame and across frames. Also, video coders take into account human factors to discard perceptually irrelevant or less important data in order to increase the overall compression ratio. As mainstream block-based coders [3] operate on blocks of pixels in a frame separately, discontinuities between block boundaries can be relevant at high compression ratios. These blocking artefacts artificially change the features within a frame, thus complicating the video-tracking task. Moreover, lossy coders discard information that, although less perceptually relevant, may be important for video tracking. For these reasons, particular care has to be taken when compression artefacts are introduced in a video prior to tracking.

An interesting link exists between video coders and algorithms that perform *tracking in the compressed domain*. As extra resources are necessary to decompress video streams, tracking in the compressed domain is preferred when computational power is limited. Compressed-domain trackers can use some by-products of coding to localise objects in the scene. For example, the phase in the frequency domain is associated with signal shifts, and possibly motion, in the spatial domain. Moreover, motion vectors produced by temporal decorrelation techniques can be used for tracking targets [4].

3.2.2   The appearance of targets

The colour appearance of objects, which is measured through the process described in the previous section, depends on three components:

- A *source* of visible electromagnetic energy

- The *object itself*, whose surface properties modulate the electromagnetic energy

- A *capturing device*, such as a camera.

The combination of these three elements makes colour difficult to manage. Understanding the physics of light, reflection and image formation is a requisite for effective feature extraction for target representation.

3.2.2.1 *Reflection models*

Reflection models describe the spatial and chromatic properties of the light reflected from a surface. Examples of reflection models are the Phong shading model and the dichromatic reflection model.

The *Phong model* [5] is an empirical local illumination model, which treats point light sources only and models three types of reflected light:

- *Ambient reflection*, which models the reflection of light arriving at the surface of the object from all directions, after being bounced around the scene in multiple reflections from all the surfaces of the scene. As we will see later, ambient reflection may have a strong influence on the perceived colour of a target.

- *Diffuse reflection*, which models the reflection from non-shiny surfaces that scatter light in all directions equally.

- *Specular reflection*, which models the reflection of light from mirror-like surfaces. The specular reflected intensity represents the highlights, the glossy reflection of objects, which is the result of the incident illumination bouncing off the surface. Highlights on the surface of a target change the local colour and gradient properties.

The intensity of the light reflected from the surface governs the appearance of a target. The colour intensity at the surface of a target is the sum of the intensities from each of the above three types of reflections.

The *dichromatic reflection model* [6] describes the light which is reflected from a point on a dielectric, non-uniform material as a linear combination (mixture) of:

- the light reflected at the material surface (*surface reflection*) and

- the light reflected from the material body (*body reflection*).

Surface reflection represents the highlights, the glossy reflection of objects. It generally has approximately the same spectral power distribution as the illumination [7]. The light which is not reflected at the surface penetrates into the material body where it is scattered and selectively absorbed. Some fraction of the light arrives again at the surface and exits the material. The light travelling through the body is increasingly absorbed at wavelengths that are characteristic for the material. The body reflection provides the characteristic target colour and reflects the light that is generally diffused equally in all directions.

3.2.2.2 Local illumination variations

Colour models that distinguish colour changes at material boundaries from colour changes due to local illumination variations, such as shading, shadows and highlights are desirable to identify targets in an image, as delimited by material boundaries. In particular, *shadows may modify the perceived target shape.*

A shadow occurs when an object partially or totally occludes direct light from a source of illumination. A shadow is visible when the position of the camera is not coincident with the light source position. Shadows can be divided into two classes:

- *Self shadows*, which occur in the portion of a target that is not illuminated by direct light;

- *Cast shadows*, which are the area projected by the target in the direction of direct light.

Note that *the spectral characteristics of the ambient light can be quite different from those of the direct light.* For example, if there is another target (or targets) casting its colour on the observed target, the latter is exposed to ambient illumination that does not have the same spectral characteristics as the incident illumination. This case is referred to as *inter-reflection.* If this additional ambient illumination influences either the lit or the shadow regions, then the colour of the two regions can be very different. This would make the task of identifying the shadow region and the lit region as part of the same observed target very complicated.

3.3   LOW-LEVEL FEATURES

This section presents an overview of colour representations, gradient and motion computation strategies. The goal is to understand how to exploit low-level features in the different stages of target representation and localisation.

3.3.1   Colour

A wide variety of mathematical representations have been proposed for the specification of colour. A colour space is a mathematical representation of our visual perceptions, and allows us to analyse and manage colour. The branch of colour science concerned with numerically specifying the colour of a physically defined visual stimulus is colorimetry.

We define colour spaces based on the Commission Internationale de l'Eclairage (CIE) colorimetry system [8, 9]. The CIE system relies on the principles of trichromacy, an empirical generalisation which sums up the experimental laws of colour-matching. The *tri-chromatic theory* states that any

Figure 3.2 Original image used for colour space comparisons. Sample frame from the PETS-2000 dataset.

colour stimulus can be matched in colour by proper amounts of three primary stimuli (e.g. red, green and blue additive colours). This theory is based on the hypothesis that the human retina has three kinds of colour sensors (cones), and that the difference in their spectral responses contributes to the sensation of colour.

The amounts of three primaries (tristimulus values of the spectrum) needed to match a unit amount of power at each wavelength of the visible spectrum is given by the *colour-matching functions* [10]. The colour-matching functions are related to the spectral sensitivities of the three cones by linear transformations. A system classifying colour according to the human visual system is defined by the CIE. This system weights the spectral power distribution of the light emitted by an object in terms of three colour matching functions. These functions are the sensitivities of a standard observer to light of different wavelengths. The weighting is performed over the visual spectrum, from around 360 nm to 830 nm in 1 nm intervals.

The above process produces three *CIE tristimulus values, XYZ*, which describe a colour (Figure 3.2 and 3.3). Chromaticity coordinates and colour spaces can be derived from these tristimulus values. If we imagine that each of the three attributes used to describe a colour are axes in a 3D space then this defines a colour space. The colours that we can perceive can be represented by the CIE system. Other colour spaces are subsets of this perceptual space.

X   Y   Z

Figure 3.3 The three components of the XYZ colour space extracted from the sample image in Figure 3.2. The visualisation of each figure is obtained by setting to zero the values of the other two components.

L a b

Figure 3.4 The three components of the Lab colour space extracted from the sample image in Figure 3.2. The figures for the two colour components were plotted by setting the luminance value to a constant equal to the mean luminance of the image.

3.3.1.1 *CIELAB* The $L^*a^*b^*$ values for XYZ tristimulus values X, Y, Z normalised to the white are given by [8]

$$L^* = 116\,(Y/Y_n)^{\frac{1}{3}} - 16 \tag{3.1}$$

$$a^* = 500\left[(X/X_n)^{\frac{1}{3}} - (Y/Y_n)^{\frac{1}{3}}\right] \tag{3.2}$$

$$b^* = 200\left[(Y/Y_n)^{\frac{1}{3}} - (Z/Z_n)^{\frac{1}{3}}\right] \tag{3.3}$$

$$C_{ab}^* = \sqrt{a^{*2} + b^{*2}} \qquad h_{ab}^* = \arctan\,(b^*/a^*) \tag{3.4}$$

where Y_n, X_n, Z_n are the tristimulus values of the reference white.[3] L^* represents lightness, a^* approximates redness-greenness, b^* approximates yellowness-blueness (Figure 3.4), C_{ab}^* chroma and h_{ab}^* hue. The L^*, a^*, and b^* coordinates are used to construct a Cartesian colour space. The L^*, C_{ab}^* and h_{ab}^* coordinates are the cylindrical representation of the same space.

3.3.1.2 *CIELUV* The $L^*u^*v^*$ values corresponding to CIE XYZ tristimulus values Y,X,Z normalised to the white are given by:

$$L^* = 116\,(Y/Y_n)^{\frac{1}{3}} - 16 \tag{3.5}$$

$$u^* = 13L^*\,(u' - u_n') \qquad v^* = 13L^*\,(v' - v_n') \tag{3.6}$$

$$C_{uv}^* = \sqrt{u^{*2} + u^{*2}} \qquad h_{uv}^* = \arctan\,(v^*/u^*) \tag{3.7}$$

where u' and v' are the chromaticity coordinates of the stimulus and u_n' and v_n' are the chromaticity coordinates of the reference white. L^* represents lightness, u^* redness-greeness, v^* yellowness-blueness, C_{uv}^* chroma and h_{uv}^* hue. As

---

[3] This is only true for white greater than 0.008856.

L   u   v

Figure 3.5 The three components of the Luv colour space extracted from the sample image in Figure 3.2. The figures for the two colour components were plotted by setting the luminance to a constant value equal to the mean luminance of the image.

in *CIELAB*, the L^*, u^* and v^* coordinates are used to construct a Cartesian colour space (Figure 3.5). The L^*, C^*_{uv} and h^*_{uv} coordinates are the cylindrical representation of the same space.

3.3.1.3 RGB The Red Green Blue (RGB) colour model is a non-linear colour space that employs a Cartesian coordinate system, with the three axes corresponding to the red, green and blue primaries (Figure 3.6). All colours lie in the unit cube subset of the 3D coordinate system. The main diagonal of the cube, with equal amounts of each primary, represents grey: black is (0,0,0) and white is (1,1,1). The transformation to convert from the RGB space to the device-independent CIE XYZ standard is given by

$$\begin{bmatrix} X \\ Y \\ Z \end{bmatrix} = \begin{bmatrix} X_r & X_g & X_b \\ Y_r & Y_g & Y_b \\ Z_r & Z_g & Z_b \end{bmatrix} \begin{bmatrix} R \\ G \\ B \end{bmatrix} \tag{3.8}$$

R   G   B

Figure 3.6 The three components of the RGB colour space extracted from the sample image in Figure 3.2. Each figure is obtained by setting to zero the values of the other two components.

where X, Y and Z are the desired CIE tristimulus values, R, G and B are the displayed RGB values and X_r, X_g and X_b are the weights applied to the monitor's RGB colours to find X, and so on. These values can be found using published specifications for the device. An example is the conversion between RGB and CIE XYZ rec601-1 (D65). In this case the matrix is given by

$$\begin{bmatrix} X \\ Y \\ Z \end{bmatrix} = \begin{bmatrix} 0.607 & 0.174 & 0.200 \\ 0.299 & 0.587 & 0.114 \\ 0.000 & 0.066 & 1.111 \end{bmatrix} \begin{bmatrix} R \\ G \\ B \end{bmatrix} \tag{3.9}$$

and the inversion by

$$\begin{bmatrix} R \\ G \\ B \end{bmatrix} = \begin{bmatrix} 1.910 & -0.532 & -0.288 \\ -0.985 & 1.999 & -0.028 \\ 0.058 & -0.118 & -0.898 \end{bmatrix} \begin{bmatrix} X \\ Y \\ Z \end{bmatrix} \tag{3.10}$$

3.3.1.4 YIQ, YUV, YCbCr   The YIQ, YUV and YCbCr colour spaces are linear and their main feature is the separation of luminance and chrominance information (Figure 3.7). Assuming that the RGB colour specification is based on the standard Phase Alternating Line (PAL) and Sequentiel Couleur à Memoire (SECAM) RGB phosphor whose CIE coordinates are: $R : x_r = 0.64$, $y_r = 0.33$, $G : x_g = 0.29$, $y_g = 0.60$, $R : x_r = 0.15$, $y_r = 0.06$, and for which the white point is: $x_n = 0.312$, $y_n = 0.329$, the mapping from RGB to YUV is defined as [11]

$$\begin{bmatrix} Y \\ U \\ V \end{bmatrix} = \begin{bmatrix} 0.299 & 0.587 & 0.114 \\ -0.147 & -0.289 & 0.437 \\ 0.615 & -0.515 & -0.100 \end{bmatrix} \begin{bmatrix} R \\ G \\ B \end{bmatrix} \tag{3.11}$$

Y   U   V

Figure 3.7   The three components of the YUV colour space extracted from the sample image in Figure 3.2. The figures for the two colour components were plotted by setting the luminance to a constant value equal to the mean luminance of the image.

H   S   V

Figure 3.8   The three components of the HSV colour space extracted from the sample image in Figure 3.2. Hue and saturation values are mapped onto the range $[0, 255]$ for visualisation. Note that, as hue is represented by an angle, white and black image levels correspond to the same hue value.

The U and V components are bipolar, that is they have positive and negative values. The inverse of the matrix in Eq. (3.11) is used for the conversion from YUV to RGB

$$\begin{bmatrix} R \\ G \\ B \end{bmatrix} = \begin{bmatrix} 1.000 & 0.000 & 1.140 \\ 1.000 & -0.394 & -0.581 \\ 1.000 & 2.028 & 0.000 \end{bmatrix} \begin{bmatrix} Y \\ U \\ V \end{bmatrix} \tag{3.12}$$

3.3.1.5 HSL   The acronym HSL (hue, saturation, lightness) represents a variety of similar colour spaces: HSI (intensity), HSV (value), HCI (chroma, colourfulness, intensity), HVC, TSD (hue, saturation, darkness), and so on. These colour spaces are non-linear transforms from RGB. For the YUV space, there is the separation of the luminance component. In particular, the HSV and the HSI spaces are considered here.

- In the *HSV model (hue, saturation, value)* (Figure 3.8), the coordinate system is cylindrical, and the subset of the space within which the model is defined is a six-sided pyramid, or hexacone. The top of the pyramid corresponds to $V = 1$, which contains the bright colours. Hue, H, is measured by the angle around the vertical axis, with red at $0°$, green at $120°$, and so on. Complementary colours are $180°$ from one another. The value of saturation, S, is a ratio ranging from 0 on the centre line (V axis) to 1 on the triangular sides of the pyramid. The pyramid is one unit high in V, with the apex at the origin. The point at the apex is black and has a V coordinate of 0. At this point the values of H and S are irrelevant. The point $S = 0$, $V = 1$ is white. Intermediate values of V for $S = 0$ (on the centre line) are the greys. When $S = 0$, the value of H is irrelevant. For example, pure red is at $H = 0$, $S = 1$, $V = 1$. Indeed, any colour with $V = 1$, $S = 1$ is akin to an artist's pure pigment used as the starting point in mixing colours. Adding white pigment corresponds to decreasing S (without changing V). Shades are created by keeping $S = 1$ and decreasing V. Tones are created by decreasing both S and V. Changing H corresponds to selecting the pure pigment with which to start.

Different transforms for the HSV space can be found in the literature. One conversion algorithm between RGB space and HSV space is given by [12]

$$H = 60h \tag{3.13}$$

$$V = \max(R, G, B) \tag{3.14}$$

$$S = \begin{cases} \dfrac{\max(R, G, B) - \min(R, G, B)}{\max(R, G, B)} & \text{if } \max(R, G, B) \neq 0 \\ 0 & \text{if } \max(R, G, B) = 0 \end{cases} \tag{3.15}$$

with h defined as

$$h = \begin{cases} \dfrac{G - B}{\delta} & \text{if } R = \max(R, G, B) \\ 2 + \dfrac{B - R}{\delta} & \text{if } G = \max(R, G, B) \\ 4 + \dfrac{R - G}{\delta} & \text{if } B = \max(R, G, B) \end{cases} \tag{3.16}$$

with $\delta = \max(R, G, B) - \min(R, G, B)$. If $H < 0$, then $H = H + 360$. $R, G, B \in [0, 1]$ and $H \in [0, 360]$ except when $S = 0$. $S, V \in [0, 1]$.

- The *HSI model* is defined as a cylindrical space, where the coordinates r, θ and z, respectively, correspond to saturation, hue, and intensity. Hue is the angle around the vertical intensity axis, with red at $0°$. The colours occur around the perimeter in the same order as in the HSV hexacone. The complement of any hue is located $180°$ further around the cylinder, and saturation is measured radially from the vertical axis, from 0 on the axis to 1 on the surface. Intensity is 0 for black and 1 for white (Figure 3.9).

As for the HSV space, different conversion algorithms from the RGB to the HSI space can be found. The transformation given in [13] and [14] is

$$I = \frac{R + G + B}{3} \qquad S = 1 - \frac{\min(R, G, B)}{I} \tag{3.17}$$

$$H = \arctan\left(\frac{\sqrt{3}(G - B)}{(R - G) + (R - B)}\right) \tag{3.18}$$

In the last equation, $\arctan(y/x)$ utilises the signs of both y and x to determine the quadrant in which the resulting angle lies. If the R, G and

H   S   I

Figure 3.9 The three components of the HSI colour space extracted from the sample image in Figure 3.2. Hue and saturation values are mapped onto the range [0, 255] for visualisation. Note that, as hue is represented by an angle, white and black image levels correspond to the same hue value.

B radial basis vectors are equally spaced on the unit circle, then the x and y components of an arbitrary point are given by

$$x = R - \frac{G+B}{2} = \frac{1}{2}[(R-G) + (R-B)] \qquad y = \frac{\sqrt{3}}{2}(G-B) \quad (3.19)$$

Another transformation can be found in [15]. Intensity, I, and saturation, S, are defined as in Eq. (3.17). If $S = 0$, H is not defined. If $S \neq 0$, then

$$H = \arccos\left(\frac{0.5((R-G) + (R-B))}{\sqrt{(R-G)^2 + (R-B)(G-B)}}\right) \qquad (3.20)$$

If $(\frac{B}{I} > \frac{G}{I})$, then $H = 360 - H$. $R, G, B \in [0, 1]$ and $H \in [0, 360]$, except if $S = 0$, when H is undefined. $S, I \in [0, 1]$.

3.3.2 Photometric colour invariants

Describing colour through features that remain the same regardless of the varying circumstances induced by the imaging process is an important requirement for some video-tracking applications. Changes in imaging conditions are related to:

- the viewing direction (a target changing its pose with respect to the camera or the camera moving with respect to the target),

- the target's surface orientation and

- illumination conditions.

These changes introduce artefacts, such as shadings, shadows and highlights.

r   g   b

Figure 3.10   The three components of the rgb colour space extracted from the sample image in Figure 3.2. The normalised $[0, 1]$ ranges are mapped back onto the $[0, 255]$ range to produce grey-level images.

Photometric invariants are functions describing the colour configuration of each image coordinate discounting one or more of the above artefacts. They are discussed in this section, assuming white illumination, neutral interface reflection and integrated white condition.

3.3.2.1   *Hue and Saturation* H and S components are insensitive to surface orientation changes, illumination direction changes and illumination intensity changes. In particular, in the colour space HSL, H is a function of the angle between the main diagonal and the colour point in the RGB space. For this reason, all possible colour points on the same surface region have to be of the same hue.

3.3.2.2   *Normalised RGB*   By dividing the R, G and B coordinates by their total sum, the r, g and b components of the rgb colour system are obtained (Figure 3.10). The transformation from RGB coordinates to this normalised colours is given by

$$r = \frac{R}{R+G+B}$$
$$g = \frac{G}{R+G+B}$$
$$b = \frac{B}{R+G+B}$$

(3.21)

This transformation projects a colour vector in the RGB cube onto a point on the unit plane. Two of the rgb values suffice to define the coordinates of the colour point in this plane. Since rgb is redundant ($b = 1 - r - g$), the preferred normalised colour space is typically formulated as [13]

$$Y = \alpha R + \beta G + \gamma B$$
$$T_1 = \frac{R}{R + G + B} \qquad (3.22)$$
$$T_2 = \frac{G}{R + G + B}$$

where α, β and γ are chosen constants such that $\alpha + \beta + \gamma = 1$. Y is interpreted as the luminance of an image pixel, and T_1 and T_2 are chromatic variables. Note that the conversion between RGB space and normalised rgb presents a singularity at the black vertex of the RGB cube. For $R = G = B = 0$, the quantities in Eq. (3.21) are undefined.

3.3.2.3 $c_1 c_2 c_3$ colour model The colour features $c_1 c_2 c_3$ are defined to denote the angles of the body reflection vector [14], [16]

$$c_1 = \arctan\left(\frac{R}{\max(G, B)}\right)$$
$$c_2 = \arctan\left(\frac{G}{\max(R, B)}\right) \qquad (3.23)$$
$$c_3 = \arctan\left(\frac{B}{\max(R, G)}\right)$$

These colour features define colours on the same linear colour cluster spanned by the body-reflection term in the RGB space and are therefore invariants for the dichromatic reflection model with white illumination.

3.3.2.4 $l_1 l_2 l_3$ colour model The colour model $l_1 l_2 l_3$ proposed in [14] is a photometric invariant for matte as well as for shiny surfaces. These colour features determine uniquely the direction of the triangular colour plane spanned by the two reflection terms. They are defined as a set of normalised colour differences. Let us define $K = (R - G)^2 + (R - B)^2 + (G - B)^2$, then

$$l_1 = \frac{(R - G)^2}{K}$$
$$l_2 = \frac{(R - B)^2}{K} \qquad (3.24)$$
$$l_3 = \frac{(G - B)^2}{K}$$

Note that when evaluating in practice the invariance properties of these colour features on real images, the assumptions of white illumination, neutral interface reflection and integrated white condition must be considered and evaluated. If they do not hold, an approximate invariance will be obtained.

3.3.3 Gradient and derivatives

Local intensity changes carry important information about the appearance of objects of interest. These changes happen *within* the object itself and at the *boundary* between the object and the background. At the boundary of the target, changes are due to different reflectance properties of the object with respect to the surrounding background. Within the target area, changes are due to different reflectance properties of object parts (e.g. hair, skin, clothes of a person), to textured areas, or to self shadows. In the following we discuss how to quantify spatial intensity changes. The formulation we will present can be easily extended to temporal changes as well.

The derivative of a 1D signal $f(y)$ in the continuous case can be defined as

$$\frac{\partial f(y)}{\partial y} = \lim_{\Delta y \to 0} \frac{f(y + \Delta y) - f(y)}{\Delta y}, \tag{3.25}$$

where Δy is a small variation of the input variable y. For a 2D signal, such as an image $I(u, v)$, we can define the partial derivatives with respect to each variable, u and v, as $\frac{\partial I(u,v)}{\partial u}$ and $\frac{\partial I(u,v)}{\partial u}$, respectively. The 2D vector $\nabla I(u, v)$ concatenating these two components is defined as

$$\nabla I(u, v) = \left[\frac{\partial I(u, v)}{\partial u} \quad \frac{\partial I(u, v)}{\partial u} \right] \tag{3.26}$$

and is referred to as the *image gradient*. The image gradient can also be represented in terms of its magnitude, $|\nabla I(u, v)|$:

$$|\nabla I(u, v)| = \sqrt{\frac{\partial I(u, v)}{\partial u}^2 + \frac{\partial I(u, v)}{\partial v}^2},$$

and orientation, $\theta_{\nabla I(u,v)}$:

$$\theta_{\nabla I(u,v)} = \arctan\left(\frac{\partial I(u, v)/\partial u}{\partial I(u, v)/\partial v} \right).$$

3.3.3.1 *Sobel operator*   Various approximations of Eq. (3.25) exist for a digital image. A widely used filter is the Sobel operator [17] that approximates the gradient components by convolving the image $I(u, v)$ with two 3×3 filters

(kernels), $S_u(u, v)$ and $S_v(u, v)$, that highlight horizontal and vertical intensity variations. The partial derivatives are computed as follows:

$$\frac{\partial I(u, v)}{\partial u} \approx \begin{bmatrix} 1 & 2 & 1 \\ 0 & 0 & 0 \\ -1 & -2 & -1 \end{bmatrix} * I(u, v) = S_u(u, v) * I(u, v) \qquad (3.27)$$

and

$$\frac{\partial I(u, v)}{\partial v} \approx \begin{bmatrix} 1 & 0 & -1 \\ 2 & 0 & -2 \\ 1 & 0 & -1 \end{bmatrix} * I(u, v) = S_v(u, v) * I(u, v), \qquad (3.28)$$

where $*$ is the discrete 2D convolution operand. As Sobel kernels are *separable*, it is computationally convenient to implement the 2D convolution as two 1D convolutions applied row- and column-wise, respectively

$$\frac{\partial I(u, v)}{\partial u} \approx \begin{bmatrix} 1 & 2 & 1 \end{bmatrix} * \begin{bmatrix} 1 \\ 0 \\ -1 \end{bmatrix} * I(u, v) \qquad (3.29)$$

and

$$\frac{\partial I(u, v)}{\partial v} \approx \begin{bmatrix} 1 & 0 & -1 \end{bmatrix} * \begin{bmatrix} 1 \\ 2 \\ 1 \end{bmatrix} * I(u, v). \qquad (3.30)$$

Similarly to the Sobel operator, the Prewitt operator uses 3×3 kernels to approximate the gradient. Moreover, other operators such as Roberts can approximate the gradient via 2×2 kernels. Due to the shape of the kernel, the Sobel operator produces a better approximation than Prewitt and Roberts operators. However implementation and computational considerations may lead to select one of the two alternative methods. Figure 3.11 shows an example of horizontal and vertical image-gradient computation using Sobel operators.

3.3.3.2 Scale selection

When targets are expected to vary their distance from the camera, it is desirable to be able to select the spatial scale of the features highlighted by the filtering process [18]. Scale selection requires blurring the original image to remove features at small scales. The blurring operation can be either performed separately before the gradient computation or in a single step. In the former case the image is first convolved with a discrete Gaussian kernel approximating

$$G_\sigma(u, v) = \frac{1}{2\pi\sigma^2} e^{-\frac{u^2 + v^2}{2\sigma^2}},$$

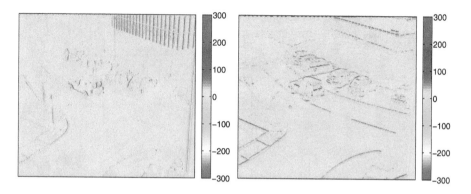

Figure 3.11   Approximation of the image gradient obtained using Sobel operators on the sample image in Figure 3.2. Left: horizontal derivative. Right: vertical derivative.

where σ^2, the Gaussian variance, is also referred to as the *scale parameter*. All the image structures that are smaller than σ will be removed from the image. The image is then filtered with a derivative operator, such as Sobel's, that is

$$\frac{\partial I(u,v)}{\partial v}(\sigma) = \frac{\partial\left(G_\sigma(u,v) * I(u,v)\right)}{\partial v} \approx S_v(u,v) * G_\sigma(u,v) * I(u,v). \quad (3.31)$$

In the latter case, using the derivative property of the convolution operator, we can express $\frac{\partial I(u,v)}{\partial v}(\sigma)$ as

$$\frac{\partial I(u,v)}{\partial v}(\sigma) = \frac{\partial\left(G_\sigma(u,v) * I(u,v)\right)}{\partial v} = \frac{\partial G_\sigma(u,v)}{\partial v} * I(u,v). \quad (3.32)$$

We can therefore convolve the image directly with the derivative of the blurring function, which in our case is the Gaussian derivative defined as

$$\frac{\partial G_\sigma(u,v)}{\partial v} = -\frac{u}{2\pi\sigma^4}e^{-\frac{u^2+v^2}{2\sigma^2}} = -\frac{u}{\sigma^2}G_\sigma(u,v).$$

Figure 3.12 shows an example of gradient approximation at different scales. As the standard deviation σ increases, the peaks on the edges become more and more blurred. Also, peaks associated to small features and noise are progressively eliminated and only filter responses associated with large structures remain visible.

As we will see in Section 3.4.2, simultaneous feature extraction at different scales may be required by some video-tracking algorithms. To this end, a Gaussian scale space [18] can be produced. The Gaussian scale space is a pyramid of images blurred at evenly spaced scales. The pyramid can be obtained by recursively applying Gaussian blurring to the image.

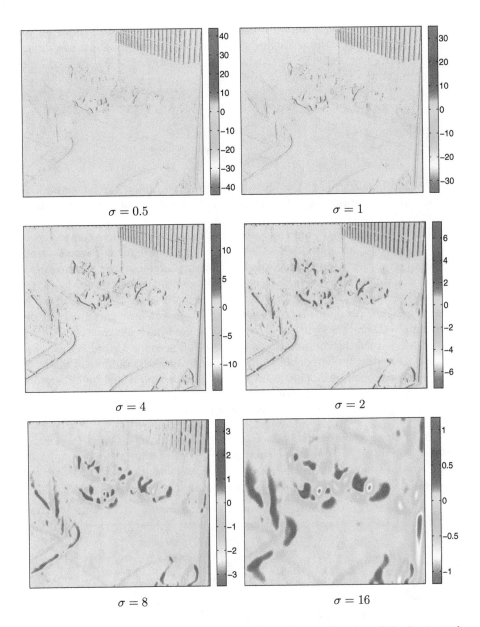

Figure 3.12 Image gradient at multiple scales. Approximation of the horizontal derivative using Gaussian derivatives with different standard deviations, σ, on the sample image in Figure 3.2. The larger σ, the larger the image features highlighted by the filter.

Let us assume that given an image resulting from Gaussian blurring the original data using a filter with scale σ^2 we want to generate a new image that is blurred at scale $R^2 * \sigma^2$, where $R^2 > 1$ defines the scale ratio. From the properties of the convolution operator.

$$G_{R*\sigma}(u, v) = G_{\Delta\sigma}(u, v) * G_\sigma(u, v) = G_{\sqrt{\Delta\sigma^2 + \sigma^2}}(u, v)$$

From this follows that

$$\Delta\sigma = \sqrt{R^2 - 1},$$

where $\Delta\sigma^2$ is the scale of the blurring operator to use on the image at scale σ^2 in order to obtain a result equivalent to blurring the original with a filter at scale $R^2 * \sigma^2$.

Note that recursively blurring using small $\Delta\sigma^2$ scales is less computationally expensive than applying to the original image a bank of filters with increasing σ^2 scales. Good approximations of the Gaussian operator require a kernel size that is proportional to the root-squared scale. On the one hand, the kernel size and the cost of convolving the original image increase with σ. On the other hand, $\Delta\sigma$ depends on the scale ratio only, and on evenly spaced scales the computational cost will not depend on the nominal scale value σ^2. Finally, to further reduce the computational cost, we can account for the loss of information caused by repeatedly blurring the image: at each doubling of the scale, neighbouring pixels will tend to store the same colour value and therefore, without loss of information, the image can be downsized by a factor of two, thus saving computations in the subsequent pyramid generation iterations.

3.3.3.3 Dealing with noise   A common problem of image-gradient computation techniques, especially at small scales, is that they tend to emphasise high spatial frequencies in the image, which can be associated with sensor-generated noise. Figure 3.13 shows an example of this undesirable effect on

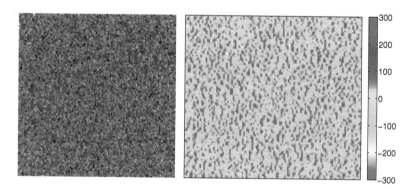

Figure 3.13   The effect of Gaussian noise on a small-matrix approximation of the image gradient. Left: synthetic Gaussian image. Right: horizontal derivative.

a synthetic image whose pixel values are generated by a Gaussian stochastic process.

To increase *robustness to noise*, the gradient can be evaluated using a least-squares estimate obtained from the structure tensor [19]

$$J(u, v) = \int \int G_\sigma (u - u', v - v') \, \nabla I(u', v')^T \nabla I(u', v') \, du' dv', \qquad (3.33)$$

where $J(.)$, the structure tensor, is a 2×2 symmetric matrix. The best local fit to the magnitude and direction of the gradient are the largest eigenvalue $\lambda_{\max}(u, v)$ of $J(u, v)$ and its associated eigenvector $\mathbf{k}_{\max}(u, v)$. Also, additional infomation about the local neighbourhood is carried by the two eigenvalues $\lambda_{\max}(u, v)$ and $\lambda_{\min}(u, v)$:

- $\lambda_{\min}(u, v) \approx 0$ in the presence of a sharp edge,

- $\lambda_{\max}(w) \approx \lambda_{\min}(w)$ if no single orientation predominates.

An *edginess measure*, $G(.)$, can be defined as [20]

$$G(w) = \sqrt[4]{\lambda_{\max}(w)^2 - \lambda_{\min}(w)^2}. \qquad (3.34)$$

$G(.)$ highlights neighbourhoods corresponding to strong straight edges and penalises regions with isotropic gradient (i.e. $\lambda_{\min}(w) \neq 0$) that are often associated with image noise.

3.3.4 Laplacian

The approximations formulated for first-order derivatives can also be extended to higher orders, such as the Laplacian:

$$\nabla^2 I(u, v) = \frac{\partial^2 I(u, v)}{\partial u^2} + \frac{\partial^2 I(u, v)}{\partial v^2},$$

i.e. the sum of the second-order derivatives, which highlights steep changes in an image. Similarly to the procedure for the Sobel filter we discussed earlier, an approximation of the Laplacian can be obtained by convolving the image with a 3×3 kernel:

$$\nabla^2 I(u, v) \approx \begin{bmatrix} 0 & -1 & 0 \\ -1 & 4 & -1 \\ 0 & -1 & 0 \end{bmatrix} * I(u, v). \qquad (3.35)$$

Figure 3.14 shows an example of a Laplacian approximation for an image. As for the first-order derivatives, the (small matrix) Laplacian operator is very sensitive to noise. As discussed before, this problem can be mitigated by

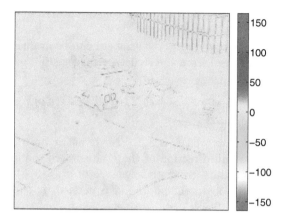

Figure 3.14 Laplacian of the sample image in Figure 3.2 obtained using a small 3×3 kernel approximation.

combining the Laplacian with a Gaussian filter. The result is the LoG operator that is defined as

$$\text{LoG}(u, v, \sigma) = \nabla^2 G_\sigma(u, v) * I(u, v), \tag{3.36}$$

with

$$\nabla^2 G_\sigma(u, v) = \frac{1}{\pi \sigma^4} \left(1 - \frac{u^2 + v^2}{2\sigma^2} \right) \exp^{-\frac{u^2 + v^2}{2\sigma^2}}. \tag{3.37}$$

Figure 3.15 shows the shape of the LoG kernel and the result of the LoG operation on a sample image. It can be seen that high variations in the filter responses are localised in the proximity of strong edges.

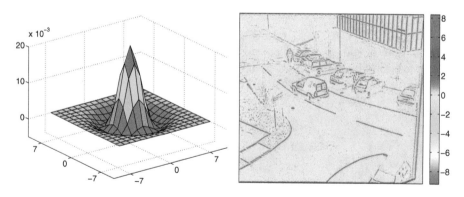

Figure 3.15 Laplacian of Gaussian (LoG) operator. Left: 3D visualisation of a LoG filter with $\sigma = 1$. Right: result after the convolution of the LoG function with the sample image in Figure 3.2.

The LoG operator can also be approximated by the difference of Gaussians (DoG) operator, which is defined as

$$\mathrm{DoG}(u, v, \sigma) = G_\sigma(u, v) * I(u, v) - G_{R\sigma}(u, v) * I(u, v) \approx \mathrm{LoG}(u, v, \sigma). \quad (3.38)$$

The scale ratio that gives the best approximation is $R \approx 1.6$. This solution is commonly applied to generate Laplacian scale spaces: starting from smaller scales, the image is iteratively convolved with a Gaussian filter with scale $\Delta\sigma$. The result is subtracted from the smoothed result generated in the previous iteration.

3.3.5   Motion

Motion is an important feature in video tracking that can be used to detect objects of interest and to favour their localisation over time. It is important to note that motion cannot be directly measured in an image sequence: a related measure is the variations of the intensity function with time. We can distinguish two kinds of motion: 2D motion and apparent motion.

- *2D motion* is the projection of the 3D real-world motion into the image plane. The distribution over the image plane of such 2D motion vectors constitutes the motion field.

- *Apparent motion* is what we consider to be motion when analysing the temporal changes of the intensity function.

From a video sequence, we can only compute apparent motion, and then we use this as an estimate of the 2D motion.

   Motion estimation aims at characterising apparent motion by a correspondence-vector field or by an optical-flow field. A *correspondence vector* describes the displacement of a pixel between two frames. The *optical flow* is the distribution of apparent velocities that can be associated with apparent motion. Estimating 2D motion through apparent motion is a very difficult task. Apparent motion is highly sensitive to noise and to variations in scene illumination. In addition to this, moving objects must be highly textured to generate apparent motion. Motion estimation faces two other problems: occlusion and aperture. The *occlusion problem* stems from the lack of a correspondence vector for covered and uncovered background. The *aperture problem* results from having a number of unknowns in the motion-estimation problem which is larger than the number of observations. Thus, assumptions are needed to obtain a unique solution. These assumptions are usually smoothness constraints on the optical-flow field to achieve continuity.

   Parametric or non-parametric representations can be used to describe the motion field. Parametric models require a segmentation of the scene and describe the motion of each region by a small set of parameters. In the case of non-parametric representation, a dense field is estimated, where each pixel is assigned a correspondence or flow vector.

The process of estimating the optical flow usually starts with a pre-filtering step to descrease the effect of noise. Then, spatio-temporal mesurements are performed to estimate the optical flow. The results are finally integrated through contraints on local or global smoothness. The spatio-temporal mesurements are based on gradients in *differential techniques* [21–23], on the outputs of band-pass filters in *phase-based techniques* [24], on analysis of the frequency domain in *energy-based techniques* [25] and on analysis of the similarity between regions in *correlation-based techniques* [26–28]. Reviews of the different techniques are given in [29,30], and comparative tests for performance evaluation are reported in [31].

3.4   MID-LEVEL FEATURES

Mapping the image onto low-level features may not be adequate to achieve a good summarisation of image content, thus reducing the effectiveness of a video tracker. A widely used solution is to analyse the video using subsets of pixels that represent relevant structures (e.g. edges and interest points) or uniform regions, whose pixels share some common properties (e.g. colour). In the next sections we will discuss how to generate these subsets for video tracking.

3.4.1   Edges

Edge detectors generally produce a binary map that highlights the presence of edges or edge elements in an image. This binary map is called an *edge map*. As discussed in Section 3.3.3, edges are often associated with high responses of first-order derivative operators. A very simple form of edge detection consists in thresholding the image-gradient magnitude. The edge map, $E(u, v)$, is therefore computed as

$$E(u, v) = \begin{cases} 1 & |\nabla I(u, v)| > T \\ 0 & \text{otherwise} \end{cases},$$

where T is an appropriate threshold that depends on the gradient computation technique and on the strength of the edges that we want to highlight. Figure 3.16(a) shows a typical result from an edge detector based on the Sobel operator.

Basic techniques such as the one just introduced present several shortcomings, mostly due to the locality of the information available during thresholding. As shown in Figure 3.16(a), detections associated with a single physical structure tend to be fragmented (i.e. the pixels labelled as edges are not spatially connected). Also, it might be difficult to select a threshold

(a)   (b)

Figure 3.16 Sample results from two different edge detectors on the image in Figure 3.2. Left: edge detection based on the Sobel operator (white pixels represent large values of the gradient magnitude). Right: edge detection based on the LoG operator (white pixels correspond to the zero-crossings of the LoG operator result).

value that allows sensitivity to weak edges while avoiding false detections caused by image noise. To exploit edge spatial-continuity, techniques like the Canny detector [32] first localise an initial edge point and then propagate the result by following the normal to the gradient direction. Another class of methods analyses variations of second- or higher-order image derivatives. A popular edge localisation technique [33] relies on detecting the zero crossings of the LoG image (Eq. 3.36). To detect horizontal and vertical zero crossings, we can threshold the absolute value of the convolution between the rows and the columns of the LoG-filtered image with a simple kernel $[-1 \quad 1]$:

$$
E(u,v) = \begin{cases} 1 & \begin{aligned} \left| \begin{bmatrix} -1 & 1 \end{bmatrix} * \mathrm{LoG}(u,v,\sigma) \right| &> T \quad \lor \\ \left| \begin{bmatrix} -1 \\ 1 \end{bmatrix} * \mathrm{LoG}(u,v,\sigma) \right| &> T \end{aligned} \\ 0 & \text{otherwise} \end{cases} .
$$

The threshold T defines the amount of gradient change that is necessary to detect an edge. A sample edge-detection result based on the LoG operator and zero-crossing detection is shown in Figure 3.16(b).

3.4.2 Interest points and interest regions

Interest-point detectors aim at selecting image features that can be accurately and reliably localised across multiple frames, when pose and illumination conditions may change. Most interest-point detectors tend to select highly distinctive local patterns such as corners, prominent edges, and regions with discriminative texture.

In this section we will describe selected interest-point detectors, ranging from the seminal Moravec detector [34] to scale-space detection approaches based on the LoG operator.

3.4.2.1 *Moravec corner detector*

The Moravec detector defines as an interest point a position in the image that is characterised by large intensity variations in multiple directions [34]. Given a small square region of fixed size W (e.g. $W = 3$ or 5) that is centred on a pixel with coordinates $c = (u, v)$, the Moravec operator computes the *intensity variations*, $\nu(c, s)$, in eight directions. These directions are up, down, left, right and the four diagonal directions. The intensity variation is computed as the sum of the squared differences between the pixel values in the region centred at c and the pixel values in the region obtained by a one pixel shift in one of the eight directions, where $s = (s_u, s_v)$ is the shift vector. This can be expressed as

$$\nu(c, s) = \sum_{|\tilde{c} - c|_1 < W} \left[I(c + \tilde{c}) - I(c + \tilde{c} + s) \right]^2, \qquad (3.39)$$

with

$$s \in \{(0, 1), (0, -1), (1, 0), (-1, 0), (1, 1), (-1, -1), (1, -1), (-1, 1)\}.$$

The measure of interest in a pixel, $\iota(c)$, is estimated as

$$\iota(c) = \min_s \nu(c, s).$$

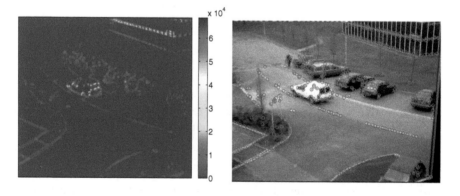

Figure 3.17 Example of interest-point detection based on the Moravec corner detector on the sample image in Figure 3.2. Left: visualisation of the Moravec interest score. Right: detected interest points (centres of the red circles).

Figure 3.18 Example of interest-point detection based on the Moravec corner detector on a toy target that undergoes rotation. As the response of the Moravec cornerness score is anisotropic, the number and location of the detected interest points may vary with the orientation of the target.

The left-hand side of Figure 3.17 shows an example of the interest score (also known as the Moravec cornerness score) on a test image using the Moravec operator.

The interest score in the image $\iota(c)$ is thresholded to find local maxima that are then associated with the interest points. The value of the threshold is chosen to eliminate spurious peaks that are due to sensor noise. The right-hand side of Figure 3.17 shows the resulting image locations that are detected after thresholding $\iota(c)$. The interest points are close to corners and strong diagonal edges.

Althouth the Moravec detector is not computationally demanding, its major drawback is that the response $\iota(c)$ is anisotropic. As a consequence, the output of the detector is not rotation invariant, thus negatively affecting the repeatability of the interest-point detection under target rotations (see Figure 3.18).

3.4.2.2 *Harris corner detector*

To favour repeatability in the detection of interest points irrespective of the target orientation, a weighted version of the differential score, $\nu(c, s)$ (Eq. 3.39), and a Taylor approximation can be used [35]. To simplify the notation without loss of generality, let us omit c by assuming that the coordinate system is centred at c (i.e. $c = (0,0)$). We can define the weighted differential score as

$$\nu(s) = \sum_{|\tilde{c}|_1 < W} w(\tilde{c}) \left[I(\tilde{c}) - I(\tilde{c} + s) \right]^2, \tag{3.40}$$

where $w(s)$ is a weighting kernel.

Unlike the Moravec detector, instead of shifting the region using a set of predefined directions, the Harris detector approximates $I(\tilde{c} + s)$ with its first-order Taylor expansion with respect to the shift s, centred at $s = (0, 0)$, that is

$$I(\tilde{c} + s) \approx I(\tilde{c}) + \nabla I(\tilde{c})^T s. \tag{3.41}$$

By substituting Eq. (3.41) into Eq. (3.40) one obtains

$$\nu(s) \approx \sum_{|\tilde{c}|_1 < W} w(\tilde{c}) \left[\nabla I(\tilde{c})^T s \right]^2.$$

Let us define the structure tensor matrix J as

$$J = \sum_{|\tilde{c}|_1 < W} w(\tilde{c}) \nabla I(\tilde{c}) \nabla I(\tilde{c})^T; \tag{3.42}$$

from this we can rewrite the approximated differential score as

$$\nu(s) \approx s^T J s.$$

As pointed out when describing the Moravec operator, corners correspond to large variations of $\nu(s)$ in multiple directions. This corresponds to a specific configuration of the eigenvalues $\lambda_{\max}$ and $\lambda_{\min}$ of the structure tensor J:

- If $\lambda_{\max} = \lambda_{\min} = 0$ (both eigenvalues are zero), then we are in a flat region of the image.
- If $\lambda_{\max} \gg \lambda_{\min}$ (there is one dominant direction), then the region corresponds to an edge.
- If $\lambda_{\min} \gg 0$ and $\lambda_{\max} \cong \lambda_{\min}$ (large and similar eigenvalues), then the region corresponds to a corner.

To avoid the computation of the eigenvalues, an alternative *cornerness score* can be used (left-hand side of Figure 3.19), defined as

$$\iota = |J| - k \operatorname{trace}(J)^2,$$

where $|.|$ denotes the determinant and k is an empirically determined constant that adjusts the level of sensitivity between *cornerness* and *edgeness*.

As a final note, the selection of an isotropic kernel (e.g. a Gaussian) for $w(.)$ generates an isotropic response to image features, thus improving the detection repeatability in case of in-plane rotations. The right-hand side of Figure 3.19 shows a sample result from the Harris corner detector that uses a Gaussian kernel with $\sigma = 1.5$ and $k = 0.07$.

3.4.2.3 Detection at multiple scales: interest regions

While the Harris corner detector improves the rotation invariance with respect to the Moravec

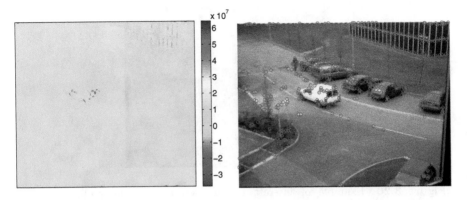

Figure 3.19 Example of interest-point detection based on the Harris corner detector on the sample image in Figure 3.2. Left: visualisation of the Harris interest score. Right: detected interest points (centres of the red circles).

detector, the scale of the features surrounding the interest point will depend on the kernel $w(.)$ and, in the case of Gaussian kernels, the variance will be linked to the scale. To favour the repeatability of the detected feature points on targets that may vary their distance from the camera (i.e. to achieve scale-invariance), we can detect features at multiple scales.

A simple solution is to produce a Gaussian scale space (see Section 3.3.3) and to apply separately the Harris (or Moravec) detector at each scale. A better approach is to detect features that are not only spatially salient, but also have a distinctive scale. Since scale information is related to the size of a region of interest around the interest point, *multi-scale interest-point detectors* are also often referred to as interest-region detectors or blob detectors.

A popular approach for detecting interest regions uses the Laplacian operator ∇^2 over a scale space [18] (Section 3.3.3). Let us assume that a LoG scale space is available, derived from a Gaussian scale space by either computing the Laplacian of all the Gaussian images or by using the LoG approximation based on DoG filters (see Section 3.3.3). We can consider the LoG scale space as a 3D function $LoG(u, v, \sigma)$ where two dimensions are the positional coordinates on the image and the third dimension spans the scale σ^2. Spatial extrema of the $LoG(u, v, \sigma)$ operator are associated to uniform dark and white blobs of sizes comparable with σ (see also Figure 3.15). Therefore, to detect interest regions that are distinctive in both space and scale, one can look for points that are extrema of the 3D function $LoG(u, v, \sigma)$. Since the magnitude of the LoG response also depends on the scale of the Gaussian filter, a common solution is to replace the Laplacian ∇^2 with the scale-normalised Laplacian $\sigma^2 \nabla^2$. In practical terms, this multi-scale approach requires comparing each 3D coordinate of the normalised function $\sigma^2 \nabla^2 (u, v, \sigma)$ with its 26 neighbours.

Further to the generic LoG-based approach, a set of post-processing steps are usually applied to improve the repeatability of the detected interest

Figure 3.20 Example of interest regions detected by the multi-scale SIFT method on the sample image in Figure 3.2. The magnitude of the arrows is proportional to the square root of the scale.

regions. For example, the interest-region detector used in the Scale Invariant Feature Transform (SIFT) [36] refines the localisation procedure by interpolating the 3D LoG function using on second-order Taylor expansion. The same approximation is also used to discard spurious regions whose magnitude $|\sigma^2 LoG(u, v, \sigma)|$ is smaller than a pre-defined threshold. Finally, poorly defined peaks associated with spatial edges in $\sigma^2 LoG(u, v, \sigma)$ are eliminated by thresholding the ratio of the eigenvalues of the LoG structure tensor.[4] Figure 3.20 shows the results of the region detector used in the SIFT approach.[5]

3.4.3   Uniform regions

Unlike video-tracking algorithms using relevant structures such as interest regions to locate a target over subsequent frames, regions whose pixels share some properties such as colour (see Section 3.3.1) or motion (see Section 3.3.5) can be defined as the basis for tracking. To generate these uniform regions, one can cluster selected low-level features within the target area [37].

Let G be the set of feature vectors, $G = \{\mathbf{g}_j : \mathbf{g}_j \in \mathbb{R}^K, j = 1, \ldots, N\}$, where N denotes the number of pixels and $\mathbf{g}_j = (g_j^1, \ldots, g_j^K)^T$. A partition can be represented as a matrix, in which the number of rows is equal to the number of clusters, C, and the number of columns is equal to N. We denote this matrix as $\mathbf{U} = \{u_{ij}\}$, where u_{ij} is the *membership value* of element $\mathbf{g}_j$ to the ith cluster, C_i. The membership value is defined as

$$u_{ij} = \begin{cases} 1 & \text{if } \mathbf{g}_j \in C_i \\ 0 & \text{otherwise.} \end{cases} \qquad (3.43)$$

---

[4] See Eq. (3.42) for the definition of the structure tensor.

[5] Results obtained using the MATLAB code available at: http://www.vlfeat.org/~vedaldi/.

Let U be the set of all $C \times N$ matrices $\mathbf{U}$, with $2 \leq C < N$. The space Π of partitions for G is the set

$$\Pi = \left\{ \mathbf{U} \in U : u_{ij} \in \{0,1\} \; \forall i,j; \; \sum_{i=1}^{C} u_{ij} = 1 \; \forall j; \; 0 < \sum_{j=1}^{N} u_{ij} < N \; \forall i \right\},$$

$$(3.44)$$

where the condition

$$\sum_{i=1}^{C} u_{ij} = 1 \qquad (3.45)$$

imposes that each object $\mathbf{g}_j$ belongs to exactly one cluster, and the condition

$$0 < \sum_{j=1}^{N} u_{ij} < N \quad \forall i \qquad (3.46)$$

means that no cluster is empty, nor do any of the clusters corresponds to the whole set G. The set Π represents the space of the solutions of the partition problem, which needs to be explored in order to obtain the final partition. As we will see in the following, the result of a clustering strategy may be either a single partition (e.g. region growing, amplitude thresholding) or a set of partitions (e.g. hierarchical clustering, pyramidal clustering).

A summary of feature-clustering approaches to generate uniform image regions is presented in Figure 3.21.

3.4.3.1 *Transition-based versus homogeneity-based clustering*   Depending on the strategy adopted to find structures in the feature space, clustering algorithms can be divided into transition-based methods (or boundary-based) and homogeneity-based (or region-based) [38, 39]. *Transition-based* methods aim to estimate the discontinuities in the feature space. They primarily use gradient information to locate boundaries [40–43]. Examples are active-contour methods (snakes) [42, 43]. Gradient information is then processed to obtain thin boundaries defining the partition. Local minima caused by noise and quantisation errors make the transition-based methods generally ill-conditioned. A small local error may in fact have significant consequences in the final result. *Homogeneity-based* methods estimate the similarity among features [44, 45]. Examples are split and merge, and region growing. The feature space is analysed on the basis of deterministic or probabilistic homogeneity criteria.

A homogeneity-based clustering algorithm may be hierarchical or partitional, based on whether the resulting classification is a nested sequence

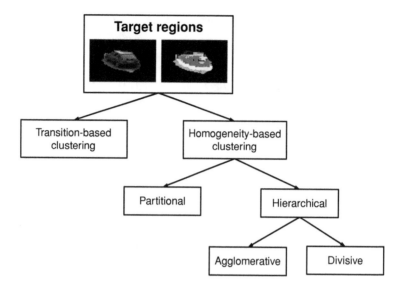

Figure 3.21 Taxonomy of feature-clustering methods for the generation of uniform image regions.

of partitions or a single partitioning. Furthermore, a homogeneity-based algorithm is agglomerative or divisive, based on whether the algorithm proceeds by gradually merging clusters into larger and larger classes or by subdividing larger clusters into smaller ones. In the following we discuss the most common homogeneity-based methods.

3.4.3.2 *Hierarchical clustering* Hierarchical algorithms provide a sequence of clustering results for a given feature space [46]. Spatially unconstrained hierarchical algorithms can be either agglomerative or divisive. *Agglomerative hierarchical algorithms* compare all the possible pairs of clusters and merge those maximising a certain similarity criterion. The process stops when all the elements of the feature space belong to one cluster.[6] *Divisive hierarchical algorithms* consider all the possible partitions at a lower level and choose the one satisfying a certain criterion. The process stops when each element of the feature space constitutes a separate cluster. Hierarchical algorithms allow one to choose the best partition level according to the specific requirements of the application, but they are computationally expensive.

A *pyramidal technique* may be seen as a spatially constrained agglomerative hierarchical algorithm. Pyramidal techniques combine local and global information to obtain partitions of the feature space. The main steps involved are

---

[6] Conditions (3.45) and (3.46) are not respected in hierarchical clustering when the extremes of the hierarchy are the image itself (all the elements of the feature space belong to one cluster) and the pixels (each element of the feature space constitute a separate cluster).

blurring, linking, root labelling and downward projection [47,48]. The *blurring* step operates in the data space, $\mathbb{Z}^M$, and creates copies of the original data at several different scales. The copies are obtained by recursively low-pass filtering the input data, and constitute the different layers of the pyramid. The *linking* step operates in the feature space, $\mathbb{R}^K$, and establishes relationships among elements at adjacent layers, according to their similarities, thus defining a tree. *Root labelling* marks the elements representing a region in the coarsest level. Finally, this labelling is propagated in the pyramid through *downward projection*, thus defining the partitions. Different levels of the pyramid correspond to different segmentations in the data space, $\mathbb{Z}^M$. This set of segmentation results allows one to select the most appropriate partition for the application at hand. Pyramidal techniques are usually not flexible, since there is no possibility of correcting the mistakes made at one level with the information of the upper levels. Furthermore, many pyramidal methods are used on quadtrees, thus producing blockiness. They are computationally complex, and produce disconnected regions.

3.4.3.3 *Partitional clustering*

Partitional algorithms generate a single partition of the visual data. Different techniques may be employed to this end, as described in the following:

- *ISODATA* or *C-Means* The partition of the feature space is determined by sequentially scanning all the feature vectors $\mathbf{g}_j$, $j = 1, \ldots, N$. The computation of the partition is based on the following steps: initialisation, classification, cluster update and cluster validation. In the *initialisation* step the number of clusters C, $2 \leq C \leq N$, is fixed and then the first partition $\mathbf{U}^{(0)}$ is generated. The successive steps are iteratively performed as follows. For each iteration $l = 0, 1, \ldots$

  1. Compute the centroid, $\mathbf{v}_i^{(l)}$, of each cluster for the partition $\mathbf{U}^{(l)}$ (*cluster update*).

  2. Associate each feature vector, $\mathbf{g}_j$, with the closest centroid to obtain $\mathbf{U}^{(l+1)}$ (*classification*).

  3. Compare the new and the previous partition, $\mathbf{U}^{(l+1)}$ and $\mathbf{U}^{(l)}$. Stop the process if the difference between the two consecutive partitions is smaller than a certain threshold, otherwise go to step 1 (*cluster validation*).

In the cluster-update step, the criterion that measures the desirability of clustering candidates may be the minimisation of an *objective function*. In such a case, the clustering process corresponds to the minimisation of a functional. Typically, local extrema of the function are defined as optimal clustering. At each iteration, the algorithm aims to evaluate the partition that minimises the functional J, expressed by

$$J(\mathbf{U}, \mathbf{v}) = \sum_{j=1}^{N} \sum_{i=1}^{C} u_{ij} \mathcal{D}(\mathbf{g}_j, \mathbf{v}_i)^2 \qquad (3.47)$$

where $\mathcal{D}(\mathbf{g}_j, \mathbf{v}_i)$ is the distance between the ith centroid $\mathbf{v}_i$ and the feature vector corresponding to the jth pixel, $\mathbf{g}_j$. For ISODATA, the distance measure is the Euclidean distance, and the measure of cluster quality is the overall within-group sum of squared errors (WGSS) between the $\mathbf{g}_j$ and the centroids $\mathbf{v}_i$ [49,50]. An interesting modification of the traditional ISODATA algorithm is to consider the membership value u_{ij} as a continuous variable, instead of a binary value. Thus, we obtain the so-called *fuzzy C-Means* algorithm [51]. Furthermore, the distance measure $\mathcal{D}(\cdot)$ may be modified, so as to influence the shape of the cluster.

- *Region growing* Region growing is based on finding similarities in the neighbourhood of a pixel to form a region. Region-growing techniques perform the following steps: initialisation, search and mark [52,53]. The *initialisation* step operates in the data space, $\mathbb{Z}^M$, by selecting a certain number of points in the image. These points constitute the seeds for the *searching* procedure which operates in the feature space, $\mathbb{R}^K$, by considering feature vectors of points in the neighbourhood of the seeds. Finally, the *mark* step selects the most similar neighbourhood according to the chosen criterion and marks it with the same label as the seed. The search and the mark steps iterate until the last point is classified.

  According to the searching technique they employ, region-growing techniques can be classified as single linkage, hybrid linkage and centroid linkage [54]. The *single linkage* works at pixel level and therefore is sensitive to noise. This technique can be improved either by introducing a pre-filtering step or by a hybrid-linkage approach. The *hybrid-linkage* approach considers the properties of a neighborhood of a pixel in the search step. Finally, in the *centroid linkage* the pixel is compared to the mean properties of the neighboring clusters.

  As general comment, region-growing methods lack global analysis and are quite sensitive to noise. Moreover, since the initialisation step is critical, the use of region-growing techniques is preferable for intermediate results and refinements.

- *Split and merge* Split-and-merge techniques *split* clusters that are not sufficiently uniform and *merge* adjacent clusters which are similar enough. Examples of split-and-merge approaches are quadtree [55, 56] and octree [57]. In general, splitting is applied until the partition is composed of regions characterised by sufficiently uniform features. Next, adjacent regions are recursively merged together in order to obtain larger uniform areas.

- *Watershed techniques* Watershed techniques are based on mathematical morphology [58]. Here, segmentation can be seen as a flooding process. Two main steps can be identified: minima detection and growing. The *minima* are detected using the spatial gradient. These minima constitute the seeds from which the clusters *grow* according to the gradient value. In this aspect, watershed techniques are similar to region-growing techniques. Watershed methods can be used to create coarse results, which are then refined with other techniques such as split and merge.

- *Amplitude thresholding* Amplitude thresholding is based on the hypothesis that regions can be distinguished based on the amplitude of their features [59,60]. Typical examples are grey-level and colour thresholding. In these cases, a histogram is constructed with the pixel intensities. A threshold is then selected either manually or by analysing the histogram representing the frequencies of occurrence of the amplitude of the features. Different strategies have been proposed for threshold selection based on scale-space analysis [61] and fuzzy logic [62]. Amplitude-thresholding methods are very effective in particular problems, such as in segmenting X-ray images or black and white printed documents, but they are not general enough due to the hypotheses on which they are based. However, effective methods to compute the thresholds exist when the statistics of the features are known. Change detection based on background subtraction is an example of amplitude thresholding that will be described in the next section.

3.5 HIGH-LEVEL FEATURES

In order to define an object of interest, one can either group mid-level features, such as interest points and regions, or can detect directly the object as a whole based on its appearance. We refer to the results of object detection as *high-level features*. High-level features can be the centroid, the whole area or the orientation of a target.

Object detection can be performed by modelling and then classifying either the background or the foreground (Figure 3.22). On the one hand, a classifier can be trained on the appearance of the *background* pixels, and then a detection is associated with each connected region (blob) of foreground pixels [37,63]. On the other hand, a set of classifiers can be trained to encode the information that is representative of a specific class of targets, the *foreground* [64]. Background-based object detectors work mostly with static cameras and are sensitive to illumination changes. Object-based detectors instead cope with camera motion, but cannot localise targets whose appearances differ from those of the training samples.

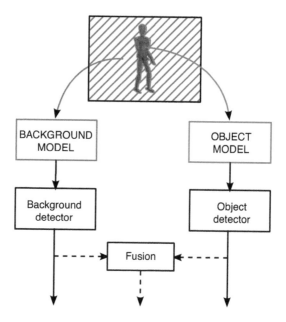

Figure 3.22 Object detection based on background and foreground models, and their fusion.

3.5.1 Background models

A variety of approaches for detecting moving objects are based on background subtraction [37, 63]. Background-subtraction algorithms use the static camera assumption to learn a model of the intesity variations of the background pixels. Foreground moving objects are detected as they cannot be explained by the (learned) background model (Figure 3.23). In production studios, background

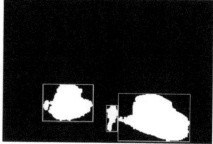

Figure 3.23 Examples of output from an object detector based on background modelling and corresponding bounding boxes representing the objects (connected components). Left: sample frame from a video surveillance scenario from i-Lids (CLEAR-2007) dataset; reproduced with permission from HOSDB. Right: object detector output.

subtraction is a popular technique whose reliability is enhanced by positioning a special type of clothing in the background [65]. The special reflective properties of this clothing facilitate the pixel classification procedure.

The estimation and the update of the *background* model can be performed by collecting statistics of the pixel values over time. Possible solutions are to use the mean and standard deviation of the pixel values [37], a mixture of Gaussians [63] or kernel-density estimation [66]. The Gaussian Mixture Model (GMM) [67–69] updates the parameters of the mixtures to cope with global illumination changes. However, objects that are stationary for a long period of time are gradually assimilated into the background model. As a consequence, the update speed is usually a trade-off between fast updates required to cope with sudden illumination changes and slow updates necessary to allow the detection of slow or stopped objects. A possible solution is to modify the learning rate in the region around a moving object depending on its speed [67]. As an alternative, a layered background can be used to maintain a record of the object appearance and keep the knowledge of the object presence. Also, edge information can help in detecting objects when they become static [68]. Once the edge structure of the background is learned, a pixel is classified as foreground by comparing its gradient vector with the gradient distribution of the background model.

Background-based methods tend to fail when in a short period of time the appearance of the background varies significantly, such as, for example, when clouds obscure the sun. The problem is that the model used to represent the typical changes is relatively simple compared to the complexity of reflections and illumination variations in a real-world scene. Another limitation of background-based detection algorithms is their intrinsic inadequacy in dealing with object interactions, such as object proximity and occlusions. In these situations, the blobs associated to two objects merge into one, thus resulting in a single-object detection. When the overlap between the objects is limited, then projection histograms can be used to split the blobs [67]. Also, motion prediction based on trajectory data can help to estimate the likelihood of an occlusion, thus allowing the association of a single blob with two objects [70].

In a number of applications, it is simpler to learn the appearance of a specific class of objects than to model the appearance of the background. This will be discussed in the next section.

3.5.2 Object models

Object detectors model the appearance of a pre-defined class of targets. The model is computed by learning representative features of the selected class (Figure 3.24).

For example, colour-based segmentation (see Section 3.4) can be used to detect faces of people. In this case, the face-detection task consists in finding pixels whose spectral characteristics lie in a specific region of the chromaticity diagram. More general object-detection approaches (e.g. not limited to colours

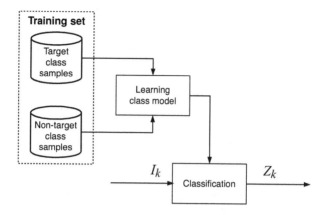

Figure 3.24 Learning a target-class model based on positive and negative samples of a target class. The class model is then used for detecting candidate targets (Z_k) in the images (I_k).

that are specific to object classes) are based on training a set of classifiers [64, 71] using images containing representative samples of the class. A target is then detected when a region in the frame generates high classifier responses to the learned features.

Edgelets [72] or Haar wavelets [73] are used in Adaboost algorithms as weak object classifiers that are combined in a cascade to form a strong object classifier [71]. Moreover, detectors trained on parts of the object can be used to deal with cases of occlusion [72].

Although more robust to changes in illumination, class-specific detectors tend to fail when other objects in the background have a similar appearance to the trained targets. A typical output of the Haar wavelets frontal-face

Figure 3.25 Sample face-detection results based on a cascade of weak classifiers [73]. The algorithm correctly detects the face but produces a false detection on the red T-shirt.

detector [73] is presented in Figure 3.25. The faces are correctly detected when are in frontal pose. However, the edges over the folds of a T-shirt produce a false detection as they are misinterpreted as the features of a face. To limit false-positive detections, object classifiers can also be used in conjunction with the results of a background-subtraction algorithm [70, 74].

3.6 SUMMARY

This chapter presented an overview of relevant features that are at the basis of the description of a target. In particular we discussed colour, motion and the gradient serving as basis for the choice of the most appropriate features for the tracking task at hand. We also discussed algorithms for the extraction of edges and interest points as well as object detectors based on the modelling of either the class of targets or the background of a scene.

In the next chapter we will describe how these features can be used to build a meaningful representation for a target.

REFERENCES

1. G.C. Holst and T.S. Lomheim. *CMOS/CCD Sensors and Camera Systems.* Bellingham, WA, SPIE Society of Photo-Optical Instrumentation Engineering, 2007.

2. OMG plc. Vicon Motion Capture System. [Hardware].

3. D. Le Gall. MPEG: a video compression standard for multimedia applications. *Communications of the ACM*, 34(4):46–58, 1991.

4. P.M. Fonseca and J. Nesvadba. Face tracking in the compressed domain. *EURASIP Journal of Applied Signal Processing*, 2006:187–187.

5. B.T. Phong. Illumination for computer generated pictures. *Communications of the ACM*, 18(6):311–317, 1975.

6. S.A. Shafer. Using color to separate reflection components. *Color Research and Application*, 10(4):210–218, 1985.

7. G.J. Klinker, S.A. Shafer and T. Kanade. A physical approach to color image understanding. *International Journal of Computer Vision*, 4:7–38, 1990.

8. M.D. Fairchild. *Color Appearance Models.* Reading, MA, Addison–Wesley Publishing Co., 1998.

9. G. Wyszecki and W.S. Stiles. *Color Science: Concepts and Methods, Quantitative Data and Formulae.* New York, John Wiley & Sons, Inc., 1982.

10. R.W.G. Hunt. *Measuring Colour.* New York, Ellis Horwood Series in Applied Science and Industrial Technology, 1987.

11. A. Watt. *Fundamentals of Three-Dimensional Computer Graphics.* Reading, MA, Addison–Wesley Publishing Co., 1989.

12. J.D. Foley, A. van Dam, S.K. Feiner and J.F. Hughes. *Computer Graphics: Priciples and Practice.* Reading, MA, Addison–Wesley Publishing Co., 1996.

13. F. Perez and C. Koch. Toward color image segmentation in analog VLSI: Algorithm and hardware. *International Journal of Computer Vision*, 12(1):17–42, 1994.

14. T. Gevers and A.W.M. Smeulders. Color-based object recognition. *Pattern Recognition*, 32:453–464, 1999.

15. R.E. Woods and R.C. Gonzalez. *Digital Image Processing*. Reading, MA, Addison–Wesley Publishing Co., 1993.

16. T. Gevers and A.W.M. Smeulders. Picktoseek: Combining color and shape invariant features for image retrieval. *IEEE Transactions on Image Processing*, 9(1):102–119, 2000.

17. R.C. Gonzalez and R.E. Woods. *Digital Image Processing (3rd Edition)*. Upper Saddle River, NJ, Prentice Hall, Inc., 2006.

18. T. Lindeberg. Scale-space theory: A basic tool for analysing structures at different scales. *Journal of Applied Statistics*, 21(2):224–270, 1994.

19. J. Bigun and G.H. Granlund. Optimal orientation detection of linear symmetry. In *Proceedings of the International Conference on Computer Vision*, Berlin, Springer-Verlag, 1987.

20. J. Bigun, T. Bigun and K. Nilsson. Recognition by symmetry derivatives and the generalized structure tensor. *IEEE Transactions on Pattern Analysis and Machine Intelligence*, 26:1590–1605, 2004.

21. H.H. Nagel. Displacement vectors derived from second-order intensity variations in image sequences. *Computer Vision, Graphic and Image Processing*, 21(1):85–117, 1983.

22. B.P. Horn and B.G. Schunck. Determining optical flow. *Artificial Intelligence*, 17(1):185–203, 1981.

23. B. Lucas and T. Kanade. An iterative image registration technique with an application to stereo vision. In *Proceedings of DARPA Image Understanding Workshop*. San Fransisco, CA, Morgan Kaufmann, 1981, 121–130.

24. D.J. Fleet, A.D. Jepson and M.R.M. Jenkin. Phase-based disparity measurement. *Computer Vision, Graphics, and Image Processing. Image Understanding*, 53(2):198–210, 1991.

25. D.J. Heeger. Optical flow using spatio-temporal filters. *International Journal of Computer Vision*, 1(6):279–302, 1988.

26. J.R. Jain and A.K Jain. Displacement measurement and its applications in intraframe image coding. *IEEE Transactions on Communications*, 29(12):1799–1808, 1981.

27. P. Anandan. A computational framework and an algorithm for the measurement of visual motion. *International Journal of Computer Vision*, 2(3):283–310, 1989.

28. F. Dufaux and F. Moscheni. Motion estimation techniques for digital TV: a review and a new contribution. *Proceedings of the IEEE*, 83(6):858–876, 1995.

29. A. Mitiche and P. Bouthemy. Computation and analysis of image motion: A synopsis of current problems and methods. *International Journal of Computer Vision*, 19(1):29–55, 1996.

30. A. Singh. *Optic Flow Computation: A Unified Perspective*. Los Alamitos, CA, IEEE Computer Society Press, 1992.

31. J.L. Barron, D.J. Fleet and S.S. Beauchemin. Performance of optical flow techniques. *International Journal of Computer Vision*, 12(1):43–77, 1994.

32. J Canny. A computational approach to edge detection. *IEEE Transactions on Pattern Analysis and Machine Intelligence*, 8(6):679–698, 1986.

33. D. Marr and E. Hildreth. Theory of edge detection. *Proceedings of the Royal Society of London Series B*, 207:187–217, 1980.

34. H. Moravec. Towards automatic visual obstacle avoidance. In *Proceedings of the 5th International Joint Conference on Artificial Intelligence*, Cambridge, MA, 1977.

35. C. Harris and M. Stephens. A combined corner and edge detection. In *Proceedings of The Fourth Alvey Vision Conference*, Manchester, UK, 1988, 147–151.

36. D.G. Lowe. Object recognition from local scale-invariant features. In *Proceedings of the International Conference on Computer Vision*, Berlin, Springer Verlag, 1999.

37. A. Cavallaro and T. Ebrahimi. Interaction between high-level and low-level image analysis for semantic video object extraction. *EURASIP Journal on Applied Signal Processing*, 6:786–797, June 2004.

38. K.S. Fu and J.K. Mui. A survey on image segmentation. *Pattern Recognition*, 13:3–16, 1981.

39. R.M. Haralick and L.G. Shapiro. Image segmentation techniques. *Computer Vision, Graphics and Image Processing*, 29:100–132, 1985.

40. T. Meier and K.N. Ngan. Automatic segmentation of moving objects for video object plane generation. *IEEE Transactions on Circuits and Systems for Video Technology*, 8(5):525–538, 1998.

41. L. Bonnaud and C. Labit. Multiple occluding objects tracking using a non-redundant boundary-based representation for image sequence interpolation after decoding. In *Proceedings of the IEEE International Conference on Image Processing*, Vol. 2, New York, IEEE, 1997.

42. M. Kass, A. Witkin and D. Terzopoulos. Snakes: active contour models. *International Journal of Computer Vision*, 1:313–331, 1998.

43. L.H. Staib and S. Duncan. Boundary finding with parametric deformable models. *IEEE Transactions on Pattern Analysis and Machine Intelligence*, 14:161–175, 1992.

44. A.A. Alatan, L. Onural, M. Wollborn, R. Mech, E. Tuncel and T. Sikora. Image sequence analysis for emerging interactive multimedia services-the european COST 211 framework. *IEEE Transactions on Circuits and Systems for Video Technology*, 8(7):802–813, 1998.

45. R. Castagno, T. Ebrahimi and M. Kunt. Video segmentation based on multiple features for interactive multimedia applications. *IEEE Transactions on Circuits and System for Video Technology*, 8(5):562–571, September 1998.

46. S. Theodoridis and K. Koutroumbas. *Pattern Recognition*. New York, Academic Press Inc., 1998.

47. H. Antonisse. Image segmentation in pyramids. *Computer Graphics and Image Processing*, 19:367–383, 1982.

48. M.R. Rezaee, P.M.J. van der Zwet, B.P.F. Lelieveldt, R.J. van der Geest and J.H.C. Reiber. A multiresolution image segmentation technique based on pyramidal segmentation and fuzzy clustering. *IEEE Transactions on Image Processing*, 9(7):1238–1248, 2000.

49. J.C. Bezdek. *Pattern Recognition with Fuzzy Objective Function Algorithm*. New York, Plenum Press, 1981.

50. A.K. Jain and R.C. Dubes. *Algorithms For Clustering Data*. Englewood Cliffs, NJ, Prentice Hall, 1988.

51. J.C. Bezdek, R. Ehrlich, and W. Full. Fcm: The fuzzy c-means clustering algorithm. *Computers & Geosciences*, 10(2-3):191–203, 1984.

52. R. Adams and L. Bischof. Seeded region growing. *IEEE Transactions on Pattern Analysis and Machine Intelligence*, 16(6):641–647, 1994.

53. O. Monga. An optimal region growing algorithm for image segmentation. *International Journal of Pattern Recognition and Artificial Intelligence*, 1(4):351–375, 1987.

54. A. Baraldi and F. Parmiggiani. Single linkage region growing algorithms based on the vector degree of match. *IEE Transactions on Geoscience and Remote Sensing*, 34(1):137–148, 1996.

55. R. Haralick and L. Shapiro. *Computer and Robot Vision*. New York, Addison Wesley, 1992.

56. A. Cavallaro, S. Marsi and G.L. Sicuranza. A motion compensation algorithm based on non linear geometric transformation and quadtree decomposition. *In Signal Analysis and Prediction*. Boston, MA, Birkhauser, 1998.

57. M. Gervautz and W. Purgathofer. *A Simple Method for Color Quantization: Octree Quantization*. New York, Springer Verlag, 1988, 219–231.

58. L. Vincent and P. Soille. Watersheds in digital spaces: an efficient algorithm based on immersion simulations. *IEEE Transactions on Pattern Analysis and Machine Intelligence*, 13:583–598, 1991.

59. W.K. Pratt. *Digital Image Processing*. New York, John Wiley & Sons, Inc., 1991.

60. A.K. Jain. *Fundamentals of Digital Image Processing*. Englewood Cliffs, NJ, Prentice Hall, 1989.

61. T. Lindeberg. *Scale-space Theory in Computer Vision*. Dordrecht, Kluwer Academic Publishers, 1994.

62. C.A. Murty and S.K. Pal. Histogram thresholding by minimizing greylevel fuzziness. *Information Science*, 60:107–135, 1992.

63. C. Stauffer and W.E.L. Grimson. Learning patterns of activity using real-time tracking. *IEEE Transactions on Pattern Analysis and Machine Intelligence*, 22(747–757), 2000.

64. B. Wu, X.V. Kuman, and R. Nevatia. Evaluation of USC human tracking system for surveillance videos. In *Proceedings of Classification of Events, Activities and Relationships (CLEAR) Workshop, Springer LNCS 4122*, Southampton, UK, April 2006, 183–189.

65. G. Thomas and O. Grau. 3D image sequence acquisition for tv & film production. In *International Symposium on 3D Data Processing Visualization and Transmission*, Los Alamitos, USA, 2002, 320.

66. A.M. Elgammal, D. Harwood and L.S. Davis. Non-parametric model for background subtraction. In *Proceedings of the European Conference on Computer Vision*. Berlin, Springer Verlag, 2000, 209–215.

67. A. Pnevmatikakis, L. Polymenakos and V. Mylonakis. The AIT outdoors tracking system for pedestrians and vehicles. In *Proceedings of Classification of Events, Activities and Relationships (CLEAR) Workshop, Springer LNCS 4122*, Southampton, UK, 2006, 171–182.

68. Y. Zhai, A. Miller P. Berkowitz, K. Shafique, A. Vartak, B. White and M. Shah. Multiple vehicle tracking in surveillance video. In *Proceedings of Classification of Events, Activities and Relationships (CLEAR) Workshop, Springer LNCS 4122*, Southampton, UK, 2006, 200–208.

69. W. Abd-Almageed and L. Davis. Robust appearance modeling for pedestrian and vehicle tracking. In *Proceedings of Classification of Events, Activities and Relationships (CLEAR) Workshop, Springer LNCS 4122*, Southampton, UK, 2006, 209–215.

70. X. Song and R. Nevatia. Robust vehicle blob tracking with split/merge handling. In *Proceedings of Classification of Events, Activities and Relationships (CLEAR) Workshop, Springer LNCS 4122*, Southampton, UK, 2006, 216–222.

71. J. Friedman, T. Hastie and R. Tibshirani. Additive logistic regression: a statistical view of boosting. *Annals of Statistics*, 28(2), 2000, 337–407.

72. B. Wu and R. Nevatia. Detection of multiple, partially occluded humans in a single image by bayesian combination of edgelet part detectors. In *Proceedings of International Conference on Computer Vision*, Washington, IEEE Computer Society, 2005, 90–97.

73. P. Viola, M. Jones and D. Snow. Detecting pedestrians using patterns of motion and appearance. In *Proceedings of the International Conference on Computer Vision Systems*. Berlin, Springer Verlag, 2003, 734–741.

74. S. Munder and D.M. Gavrila. An experimental study on pedestrian classification. *IEEE Transactions on Pattern Analysis and Machine Intelligence*, 28(11):1863–1868, 2006.

4

TARGET REPRESENTATION

4.1 INTRODUCTION

A target representation is a model of the object of interest that is used by a tracking algorithm. This model includes information about the shape and the appearance of the target. The model for a specific target can be computed in different ways:

- It can be defined a priori
- It can be a snapshot of the target
- It can be learned from a set of training samples.

In this chapter we review relevant target-representation strategies and we study in detail how to define a target in terms of its shape and its appearance. The shape and the appearance information of the model can be encoded at different levels of resolution or rigidity. For example, one can use a bounding box or a deformable contour to approximate the shape of the target, whereas the appearance information can be encoded as a probability density function of some appearance features computed within the target area.

Video Tracking: Theory and Practice. Emilio Maggio and Andrea Cavallaro
© 2011 John Wiley & Sons, Ltd

Illumination changes, clutter, target interactions and occlusions generate uncertainties in the observations. These uncertainties have to be taken into account when designing a tracker and in particular when defining how to represent a target, as discussed in the following sections.

4.2 SHAPE REPRESENTATION

As mentioned in Section 1.3, there are several ways to represent (approximate) the shape of a target. We divide the shape approximations into three groups: basic, articulated and deformable representations (Figure 4.1). These classes are discussed in the next sections.

4.2.1 Basic models

4.2.1.1 Point approximation   The simplest approach is to approximate the object shape with a *single point* on the image plane that usually corresponds to the object centroid (left side of Figure 4.1). This choice is common with trackers based on object detectors (see Chapter 7), as the appearance and size of the object in the image is accounted for at the detection stage. A large body of literature on point-tracking algorithms exists as this problem

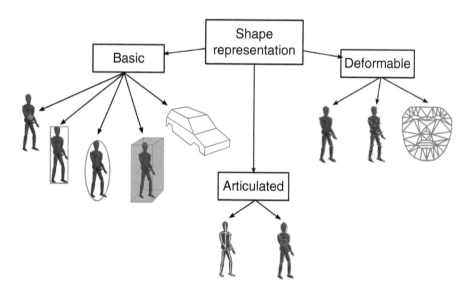

Figure 4.1 Summary of shape approximations for video tracking. Left: basic representations (single point, rectangle, ellipse, cuboid, wire-frame). Centre: articulated representations based on connected points or regions. Right: deformable contours and point-distribution models.

has been largely investigated by the radar community, where the point approximation is dominant [1]. Point trackers have important limitations when a target is occluded as they do not provide an estimate of the target size. For this reason, it is not possible to detect overlaps on the image plane using single-point-based representations.

4.2.1.2 Area approximation   *Area*-based representations bound the target with a simple shape like a rectangle or an ellipse [2, 3] (left side of Figure 4.1). The tracking algorithm relies then on information provided by the entire target area. Examples of such information are motion, colour and gradient [4]. Tracking is performed by estimating the parameters describing possible *transformations of these approximated shapes*. For example, in the case of the ellipse, the position of the centroid, the length of the axes and the rotation angle are part of the target state x_k [2]. Affine parameters instead are often used when a parallelogram bounds the target [5].

4.2.1.3 Volume approximation   Although area trackers combined with appropriate appearance models can infer occlusions (see Section 4.3.3), the lack of depth information limits the effectiveness of these shape approximations to simple tracking scenes.

When possible, tracking the spatial *volume* occupied by the target facilitates handling multiple objects occluding each other because, given camera calibration information, the relative position with respect to the camera can be estimated. One can bound the target volume using a cuboid or an ellipsoid (left side of Figure 4.1). This solution is often used when the actual 3D shape of the object is not known a priori. The volume can be estimated using either monocular vision under strong assumptions on the target movement (e.g. the target moves on a plane) or multiple calibrated cameras with overlapping views. Less stringent assumptions are required when the object is rigid and its 3D shape (or a good approximation of it) is known a priori. For example, 3D models like the one shown on the left side of Figure 4.1 have been largely used [6]. The major drawback of these methods is their lack of generality. This prevents tracking objects that are not in the database of available 3D shapes. For example, the method in [6] is aimed at highway traffic surveillance, thus the models in the database correspond to vehicles.

4.2.2 Articulated models

Articulated models approximate the shape of the target by combining a set of rigid models (i.e. points, areas or volumes) based on topological connections and motion constraints [7]. The motion constraints between connected rigid models are defined by kinematic joints to form a set of kinematic chains. For example, simplified versions of the topology of the human skeleton can be used for full-body tracking. In this case the state is composed of the global 3D position of the target augmented with a set of parameters representing the joint angles between bones (middle of Figure 4.1). Examples of articulated models are those used in motion-capture applications (see Chapter 2). In *marker-based* motion capture, actors sensorised with markers are tracked in 3D using calibrated cameras. The marker model uses a point approximation where one or more markers are associated with a limb. The relative motion between the groups of markers is represented by the articulated model (centre of Figure 4.1). By solving the inverse kinematic problem, the skeletal model is fit to the tracked marker data. In *marker-less* motion capture, the point model is substituted with a coarse region or volume-based approximation or with a more complex surface model [8].

4.2.3   Deformable models

Shape rigidity or kinematic assumptions as described above cannot be enforced for all target classes. In fact, prior information on the object shape may not be available, or the object of interest may undergo deformations that are not well modelled by canonical joints. To represent these types of targets fluid models, contours or point distribution models can be used.

4.2.3.1   Fluid models   Instead of tracking the entire target area, homogeneous regions or interest points can be identified on the object and then used to track its parts. *No explicit motion constraint* between the different parts is enforced. For example, in [9] the parts to be tracked are the detectable corners of the object. Independently tracking the parts results in stable object tracks even in cases of partial occlusions. However, the problem of grouping the points to determine which of them belong to the same object is a major drawback of these approaches.

4.2.3.2   Contours   A *contour* provides a more accurate description of the target shape [10, 11] (right side of Figure 4.1) than the other representations described so far. Contour-based trackers generally represent the shape using a set of control points positioned along the contour. The state is the concatenation of the coordinates of these points [10]. Given the high dimensionality of this representation, additional constraints are used to facilitate the contour optimisation procedure and to avoid degenerate solutions. In general, regular and smooth curves are preferred to high-curvature contours [12]. Alternatively, if prior knowledge of the target shape is available, one can learn a model of valid contour variations. As the data is mapped onto a space that encodes the major variation modes only, this solution reduces the dimensionality of the shape parameterisation [13].

4.2.3.3   Point distribution models   A number of applications require knowledge of the deformations inside the object boundary. An example is face tracking, where estimating the eyebrow and mouth positions can be useful to infer human emotions (Figure 4.2). A 2D deformable model (Figure 4.2(a)) is

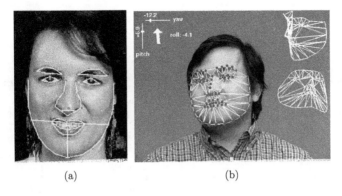

(a)   (b)

Figure 4.2 Deformable point-distribution models for face tracking: 2D (a) and 3D (b) deformable models. (a) Image courtesy of Tim Cootes (University of Manchester); (b) IEEE © [14].

parameterised by a contour as well as internal control points [15]. The control points are joined to form triangular mesh models. The mesh forms a tessellation of the target shape into triangular patches that are used for appearance representation. This representation is based on the assumption that the variations of appearance of the target can be defined by a piece-wise affine transformation. When sufficient information is available it is also possible to use image information to track a piece-wise 3D shape [14], as shown in Figure 4.2(b). In both cases a model of the plausible deformation is learned from training data.

4.3 APPEARANCE REPRESENTATION

The appearance representation is a model of the expected projection of the object appearance onto the image plane. Unlike the models used by object detectors described in Section 3.5, the representation of the appearance of a target may be *specific to a single object* and does not necessarily generalise across objects of the same class. For example, a target representation can encode the colour of a vehicle, which may be different from the colours of other vehicles in the video. An appearance model is either extracted from one or more images, or provided externally (Figure 4.3).

An appearance representation is usually paired with a function that, given the image, estimates the likelihood (or score function) of an object being in a particular state. A desirable property for this function is its *smoothness* with respect to small variations in the target position and size. Smoothness facilitates the localisation task of the tracker as it reduces the probability of the estimator converging to sub-optimal solutions.

We discuss below different appearance representations and their associated score functions.

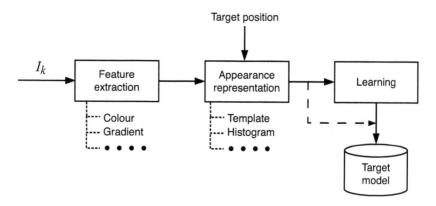

Figure 4.3 Learning a target model based on one or multiple target samples, and on a selected appearance representation.

4.3.1 Template

A common target representation is the template (Figure 4.4(b)), which encodes the positional information of colour or grey-level values for all the pixels within the target area [5, 16].

Let $\mathcal{A}$ be a transformation that, given the state x, maps a pixel position w in the coordinate system of the template $I_T(.)$ onto the coordinate system of the input image $I(.)$. Common score functions for the template are the L_1 norm:

$$d_{L_1}(x, I_T, I) = \sum_{w \in I_T} |I(\mathcal{A}(w, x)) - I_T(w)|, \tag{4.1}$$

the L_2 norm:

$$d_{L_2}(x, I_T, I) = \sum_{w \in I_T} (I(\mathcal{A}(w, x)) - I_T(w))^2, \tag{4.2}$$

or the normalised cross-correlation coefficient:

$$d_C(x, I_T, I) = 1 - \frac{1}{|I| - 1} \sum_{w \in I_T} \frac{(I(\mathcal{A}(w, x)) - \bar{I})(I_T(w) - \bar{I}_T)}{\sigma_I \sigma_{I_T}}. \tag{4.3}$$

The bar sign and σ indicate the mean and variances of the pixel values for the template $I_T(.)$ and the candidate image region defined by the state x.

The transformation $\mathcal{A}$ depends on the meaning of the parameters included in the state. Typically $\mathcal{A}$ can include translation, scaling or more complex

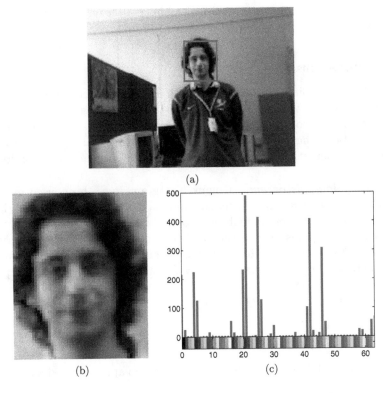

(a)

(b)

(c)

Figure 4.4 Examples of target representations: (a) a sample face target (red bounding box) and the corresponding (b) colour template and (c) colour histogram (encoded with 64 bins).

affine and perspective transformations. Note that the transformation $\mathcal{A}$ may lead to non-integer pixel coordinates. In this case it is necessary to apply a suitable image-interpolation technique, such as nearest-neighbour, bilinear or bicubic [17].

Although template-score computations are simple and fast, the intensity values stored in the template may become non-representative of the object appearance in the presence of noise, partial occlusions and unmodelled transformations. For example, unless a 3D model is available, it is difficult to predict the template appearance when the object undergoes out-of-(image)-plane rotations. In fact, because of the rigidity of this representation, templates are sensitive to noise, the score functions might be not sufficiently smooth and some values within the template might be invalid in the case of occlusions.

Several methods are available to increase the robustness of template-based representations. For example, one can average the values across neighbouring pixels or can substitute these values with image-filter responses that highlight light invariant features like edges, corners and textures [18].

4.3.2 Histograms

Instead of describing the target with the complete pixel information (position and intensity values), one can extract from the image a transformation-invariant description. In this section we describe a series of histogram-based representations that can characterise the target area and we will introduce colour and orientation histograms encoding local and global information.

4.3.2.1 *Colour histograms* Colour histograms have been used as target models for their invariance to scaling and rotation, robustness to partial occlusions, data reduction and efficient computation [2, 3, 19, 10].

Colour histograms encode the statistical distribution of the pixel values (Figure 4.4(c)). As each histogram bin contains global information, this representation, although less descriptive, is more invariant in case of partial occlusions and pose changes.

The normalised and weighted colour distribution

$$r_k(x) = \{r_{k,j}(x)\}_{j=1}^{N_b}$$

of a target candidate x over the image I_k, can be approximated as [2]

$$r_{k,j}(x) = C_h \sum_{i=1}^{n_h} \kappa \left(\left\| \frac{y - w_i}{h} \right\|^2 \right) \delta \left[b(I_k, w_i) - j \right], \qquad (4.4)$$

where

- $\kappa(.)$ is the kernel profile with bandwidth h that usually weights more the pixels near the centre of the bounding ellipse approximating the target area (Eq. 1.3)

- $\{w_i\}_{i=i,...n_h}$ are the coordinates of the pixels inside the ellipse mapped in a circle with radius h and normalised by e and θ

- $b(w_i, I_k)$ associates the pixel in position w_i with the histogram bin

- C_h is a normalisation factor defined so that the bin values add up to one.

The model-candidate matching can be assessed by using the L_1 or L_2 norm or an error measure designed to compare two probability distributions [20]. For example, in [2] the distance between the model histogram $r_{\mathcal{M}}$ and the candidate histogram $r_k(x)$ is defined as

$$d[r_k(x), r_{\mathcal{M}}] = \sqrt{1 - \rho[r_k(x), r_{\mathcal{M}}]}, \qquad (4.5)$$

Figure 4.5 Example of lost track using colour histograms as target representation. The colour distribution of the box on the bottom-left of the image is similar to the colour distribution of the face, thus misleading the tracker while the head is turning. IEEE © [21].

where ρ is the Bhattacharyya coefficient [20]

$$\rho\left[r, r_{\mathcal{M}}\right] = \sum_{j=1}^{N_b} \sqrt{r_j \cdot r_{\mathcal{M}j}}. \tag{4.6}$$

It is important to observe that a target model defined by colour histograms only can be misled:

- by changes in scene illumination
- by out-of-plane object rotations
- by background clutter.

This leads to poor performance when the object undergoes pose changes, as the similarity measure in Eq. (4.5) produces unreliable candidate-model matches.

Figure 4.5 shows an example of a *lost track* due to the use of colour histograms only: the tracker is uncertain about the position of the target as the box on the bottom-left area of the image is a good candidate target region when the head is turning away from the camera.

To address this problem *gradient* information can be used to complement colour information [19, 22], as described below.

4.3.2.2 *Orientation histograms*

The same representation used to encode the colour information in the target area can be applied to other low-level features, such as the image gradient [23]. The image gradient, as discussed in the previous chapter, highlights strong edges in the image that are usually associated with the borders of a target.

To use the image gradient in the target representation, one can compute the projection of the gradient perpendicular to the target border [19] or the edge density near the border using a binary Laplacian map [22]. However, these forms of representation discard edge information inside a target [19, 22] that may help recover the target position, especially when the target appearance is highly characterised by a few texture patterns. More detailed edge information can be obtained from the histogram of the gradient orientation [23], also known as orientation histograms.

The computation of orientation histograms is conceptually similar to that of colour histograms (Eq. 4.4). For each pixel w_i in the region of interest, the magnitude of the gradient

$$|\nabla I\left(w_i\right)|$$

is accumulated on the bin corresponding to its orientation

$$\psi\left(w_i\right).$$

This type of representation has seen wide application in local-feature descriptors (SIFT) [24].

The orientation histogram can be based on the structure tensor (Eq. 3.33) and is obtained by quantising the range $[-\pi/2, \pi/2]$ into $N_{o,b}$ bins. For each position w, the value of $G(w)$ is cumulated in the bin corresponding to the orientation $\phi(w)$ of the vector $\mathbf{k}_{\max}(w)$. In general, it is desirable to have a representation that is invariant to target rotations and scale variations:

- *Invariance to rotation* is achieved by shifting the coefficients of the histogram according to θ, the target rotation associated with a candidate state x.

- *Scale invariance* is achieved by generating a derivative scale space. The orientation histogram of an ellipse with major axis h is then computed using the scale-space-related level σ closest to h/R, where R is a constant that determines the level of detail.

4.3.2.3 *Structural histograms*

The descriptiveness of histograms is limited by the lack of spatial information, which makes it difficult to discriminate targets with similar colour or gradient distributions.

For example, Figure 4.6(b) visualises the values of the model-candidate coefficients of Eq. (4.6) when the target is the pedestrian and colour histograms are used. The area of the image where ρ is large (red area) is not concentrated near the target centroid. The white car presents a similar colour distribution and returns high coefficient values as well. In this case, the lack of discriminative power is likely to produce a tracking error, as shown in Figure 4.7.

To reduce this problem, *spatial information* can be introduced by associating with each bin of the histogram the first two spatial moments of the pixel coordinates of the corresponding colour (spatiograms) [26]. A spatiogram can

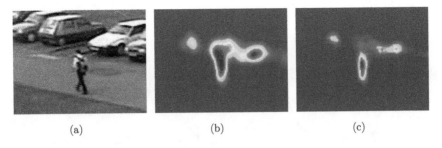

(a)   (b)   (c)

Figure 4.6 Comparison of model-candidate similarity (i.e. Bhattacharyya coefficient ρ) using a single colour histogram and the structural histogram: (a) sample frame; (b) ρ for single histogram; (c) ρ for structural histogram. Red indicates higher model-candidate similarity. The structural histogram is a more discriminative target representation than the classic single colour histogram. Reproduced with permission of Elsevier [25].

be interpreted as a GMM where each component is a 2D (spatial) Gaussian associated to and weighted by a bin of the histogram.

Alternatively, one can split the target area into regions with a regular grid and compute multiple histograms (one per region) [19, 24, 27, 28]. In this case the target is represented by a vector concatenating the values of the histograms of each region. For example, a multi-region orientation histogram can be used, where the regions are the four sectors of the bounding ellipse (Figure 4.8(b)) [28].

Another multi-region representation [27] divides a target into two non-overlapping areas (top and bottom regions). This solution is effective for a specific application (i.e. tracking ice-hockey players), as it generally corresponds to the shirt and the trousers, but it is not necessarily effective on a generic target.

(a)   (b)   (c)

Figure 4.7 Failure modality of the target representation based on a single colour histogram. The colour distribution of the background is similar to the target colour distribution. The ambiguity of the information produces a lost track. Reproduced with permission of Elsevier [25].

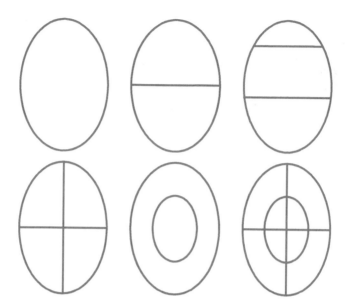

Figure 4.8 Target shape subdivisions used to generate spatially sensitive appearance (e.g. histogram-based) representations.

For elliptic tracking, a *semi-overlapping subdivision* of the target that incorporates both global and local target information in a single model can be used (Figure 4.8) [28]. We will refer to this representation as *structural histogram*. To improve the sensitivity to rotations and anisotropic scale changes, while maintaining the robustness and flexibility typical of single colour histograms the following representation can be used:

- The first histogram is calculated over the whole target.

- To account for *rotations*, four regions are then obtained from the partition created by the two axes.

- To account for *scale* changes, the inner and outer regions of a concentric ellipse with same eccentricity, but half axis size of the whole ellipse, are considered.

Note that by encoding the local distribution of the colours, this subdivision avoids representation ambiguities when the object is close to circular, and the ambiguity is now restricted to the case of circular objects with circular symmetry of the colours.

Equation (4.6) can be extended to the semi-overlapping target partition [21] as

$$\rho\left[r_k(x), r_{\mathcal{M}}\right] = \frac{\sum_{j=1}^{N_b} \rho\left[r_k^{(j)}(x), r_{\mathcal{M}}^{(j)}\right]}{N_b}, \tag{4.7}$$

Figure 4.9 Sample results on pedestrian (PETS dataset) and face tracking using different target representations. Left: single (colour) histogram. Right: structural (colour) histogram. Reproduced with permission of Elsevier [25].

where N is the number of regions, $r_k^{(j)}$ and $r_\mathcal{M}^{(j)}$ are the model and candidate histograms with N_b bins calculated on the jth target region. The model-candidate distance is now computed as in Eq. (4.5) using Eq. (4.7).

Figure 4.6 visualises the values of the model-candidate coefficients of Eq. (4.6) and Eq. (4.7) for the problem described in Figure 4.7 using colour histograms. It can be seen that the area of the image where ρ is large (red area) is narrower for the structural representation (Figure 4.6(c)), thus reducing the probability of attraction to a false target (*clutter*). Figure 4.9 shows a visual comparison between using a single histogram and the structural histogram. Numerical comparisons and further visual examples are available in Section A.1 in the Appendix.

4.3.3 Coping with appearance changes

As discussed in Section 1.2.1, the dynamic behaviour of objects as well as changing scene conditions generate variations in the appearance of the target. To cope with this challenge, model update strategies can be used (Figure 4.10).

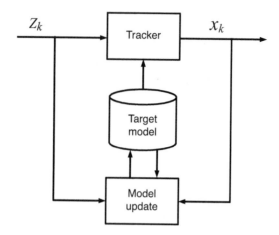

Figure 4.10   Updating the appearance representation. Image measurements and the output of the tracker can be used to adapt the appearance representation of the target to the current scene conditions.

A generic solution is to model each parameter in the target representation as a mixture of Gaussians that is *updated* over time [29]. Alternatively, the representation can *evolve* based on a modified Kalman filter [30] (see Section 5.2.2.1). It is important to note that a risk associated to updating the target model online by using the output of the tracker is *model drifting*. Model drifting, the gradual impoverishment of information in the model, is due to the amplification of the tracking error caused by the update feedback loop. To reduce this problem, the update strategy can include a contribution from the initial model [31].

When training data is available, one can *learn* a model of typical appearance changes [32, 15], for example for targets such as faces. Given the model, it is possible to predict appearance changes due to different views and expressions. In this case, a high-dimensional appearance space is mapped via Principal Components Analysis (PCA) onto a lower-dimensional space where the dominant appearance variations are well represented. Also, to further reduce the dimensionality, one can merge the appearance with the shape parameters using PCA and then simultaneously track appearance and shape [15].

4.4 SUMMARY

This chapter presented an overview of techniques for representing a target for video tracking. We discussed the advantages and disadvantages of various object representations, such as the template, which simply stores luminance (or colour) values, and several histogram representations that store information on the feature distribution in the target area. Finally we discussed approaches to approximate the shape of a target using rigid, articulated and deformable models.

Based on the information encoded in target representations, in the next chapter we address the target localisation problem.

REFERENCES

1. Y. Bar-Shalom and T.E. Fortmann. *Tracking and Data Association*. New York, Academic Press, 1988.

2. D. Comaniciu, V. Ramesh and P. Meer. Kernel-based object tracking. *IEEE Transaction on Pattern Analysis and Machine Intelligence*, 25(5):564–577, 2003.

3. P. Perez, C. Hue, J. Vermaak and M. Gangnet. Color-based probabilistic tracking. In *Proceedings of the European Conference on Computer Vision*, Vol. 1, Copenhagen, Denmark, May-June 2002, 661–675.

4. A. Cavallaro, O. Steiger and T. Ebrahimi. Tracking video objects in cluttered background. *IEEE Transactions on Circuits and Systems for Video Technology*, 15(4):575–584, 2005.

5. J. Shi and C. Tomasi. Good features to track. In *Proceedings of the IEEE Conference on Computer Vision and Pattern Recognition*, Seattle, USA, 1994, 593–600.

6. D. Koller, K. Danilidis and H.-H. Nagel. Model-based object tracking in monocular image sequences of road traffic scenes. *International Journal of Computer Vision*, 10(3):257–281, 1993.

7. A. Sundaresan and R. Chellappa. Multi-camera tracking of articulated human motion using shape and motion cues. *IEEE Transactions on Image Processing*, 18(9):2114–2126, 2009.

8. D. Anguelov, P. Srinivasan, D. Koller, S. Thrun, J. Rodgers and J. Davis. Scape: shape completion and animation of people. *ACM Transactions on Graphics*, 24(3):408–416, 2005.

9. D. Beymer, P. McLauchlan, B. Coifman and J. Malik. A real-time computer vision system for measuring traffic parameters. In *Proceedings of Computer Vision and Pattern Recognition (CVPR)*, San Juan, Puerto Rico, 1997, 495–501.

10. M. Isard and A. Blake. CONDENSATION – conditional density propagation for visual tracking. *International Journal of Computer Vision*, 29(1):5–28, 1998.

11. Y. Wu, G. Hua and T. Yu. Switching observation models for contour tracking in clutter. In *Proceedings of the IEEE Conference on Computer Vision and Pattern Recognition*, Madison, WI, 2003, 295–304.

12. M. Kass, A. Witkin and D. Terzopoulos. Snakes: Active contour models. *International Journal of Computer Vision*, 1(4):321–331, 1988.

13. T.F. Cootes, C.J. Taylor, D.H. Cooper and J. Graham. Active shape models – their training and application. *Computer Vision and Image Understanding*, 61:38–59, 1995.

14. J. Xiao, S. Baker, I. Matthews and T. Kanade. Image on 3D AAM from real-time combined 2D+3D active appearance models. In *Proceedings of the IEEE Conference on Computer Vision and Pattern Recognition*, Washington, DC, 2004, 535–542.

15. T.F. Cootes, G.J. Edwards and C.J. Taylor. Active appearance models. In *Proceedings of the European Conference on Computer Vision*, London, UK, 1998, 484–498.

16. B.D. Lucas and T. Kanade. An iterative image registration technique with an application to stereo vision. In *Proceedings of the 7th International Joint Conference on Artificial Intelligence (IJCAI '81)*, Vancouver, BC, Canada, 1981, 674–679.

17. R.C. Gonzalez and R.E. Woods. *Digital Image Processing* (3rd Edition). Upper Saddle River, NJ, Prentice-Hall, Inc., 2006.

18. A.D. Jepson, D.J. Fleet and T. El-Maraghi. Robust online appearance models for visual tracking. *IEEE Transactions on Pattern Analysis and Machine Intelligence*, 25(10):1296–1311, 2003.

19. S. Birchfield. Elliptical head tracking using intensity gradients and color histograms. In *Proceedings of the IEEE Conference on Computer Vision and Pattern Recognition*, Santa Barbara, CA, 1998, 232–237.

20. A. Bhattacharyya. On a measure of divergence between two statistical populations defined by probability distributions. *Bulletin of the Calcutta Mathmatical Society*, 35:99–109, 1943.

21. E. Maggio, F. Smeraldi and A. Cavallaro. Adaptive multifeature tracking in a particle filtering framework. *IEEE Transactions on Circuits and Systems for Video Technology*, 17(10):1348–1359, 2007.

22. T.L. Liu and H.T. Chen. Real-time tracking using trust-region methods. *IEEE Transactions on Pattern Analysis and Machine Intelligence*, 26(3):397–402, 2004.

23. W.T. Freeman and M. Roth. Orientation histograms for hand gesture recognition. In *Proceedings of the Workshop on Automatic Face and Gesture Recognition*, Zurich, Switzerland, 1995, 296–301.

24. D.G. Lowe. Object recognition from local scale-invariant features. In *Proceedings of the International Conference on Computer Vision*, Corfu, Greece, 1999, 1150–1157.

25. E. Maggio and A. Cavallaro. Accurate appearance-based bayesian tracking for maneuvering targets. *Computer Vision and Image Understanding*, 113(4):544–555, 2009.

26. S. Birchfield and S. Rangarajan. Spatiograms versus histograms for region-based tracking. In *Proceedings of the IEEE Conference on Computer Vision and Pattern Recognition*, Vol. 2, San Diego, CA, 2005, 1158–1163.

27. K. Okuma, A. Taleghani, N. de Freitas, J.J. Little and D.G. Lowe. A boosted particle filter: Multitarget detection and tracking. In *Proceedings of the European Conference on Computer Vision*, Prague, Czech Republic, 2004, 28–39.

28. E. Maggio and A. Cavallaro. Multi-part target representation for colour tracking. In *Proceedings of the IEEE International Conference on Image Processing*, Vol. 1, Genoa, Italy, 2005, 729–732.

29. S. Zhou, R. Chellappa and B. Moghaddam. Visual tracking and recognition using appearance-based modeling in particle filters. *IEEE Transactions on Image Processing*, 13:491–1506, 2004.

30. H.T. Nguyen and A.W.M. Smeulders. Fast occluded object tracking by a robust appearance filter. *IEEE Transactions on Pattern Analysis and Machine Intelligence*, 26(8):1099–1104, 2004.

31. I. Matthews, T. Ishikawa and S. Baker. The template update problem. *IEEE Transactions on Pattern Analysis and Machine Intelligence*, 26(6):810–815, 2004.

32. M. Black and A. Jepson. Eigen–tracking: Robust matching and tracking of articulated objects using a view–based representation. *International Journal of Computer Vision*, 36(2):63–84, 1998.

5

LOCALISATION

5.1 INTRODUCTION

In this chapter we will discuss how to localise a target over time, given its initial position. After initialisation, the localisation step of a video tracker recursively estimates the state x_k, given the features extracted from the video frames and the previous state estimates $x_{1:k-1}$.

We can classify localisation methods into two major classes:

- *Single-hypothesis localisation* (SHL) methods, where only one track candidate estimate is evaluated at any time

- *Multiple-hypothesis localisation* (MHL) methods, where multiple track candidates are evaluated simultaneously. The ability to propagate multiple hypotheses can improve the performance of a tracker.

The state estimate is based on the assumption that the position, shape and (optionally) the appearance of a target change smoothly over time. In other words, the tracker enforces consistency in the trajectory. The degree of

Video Tracking: Theory and Practice. Emilio Maggio and Andrea Cavallaro
© 2011 John Wiley & Sons, Ltd

enforcement varies from tracker to tracker. The various localisation strategies for video tracking will be discussed in detail in the following sections.

5.2 SINGLE-HYPOTHESIS METHODS

The optimal target state estimates (in a least square sense) can be obtained algebraically under very stringent assumptions for the relationship between the feature space E_o and the state space E_s. Often when the localisation algorithm follows a target detection step (i.e. the feature space is the detection space), the optimal estimate in the least-squares sense can be computed using the Kalman filter [1]. However, as in most cases the relationship between features and state does not fulfill the optimality conditions, an approximated solution is necessary. An example are the gradient-based methods that will be discussed next.

5.2.1 Gradient-based trackers

Gradient-based trackers use image features to steer the state candidate (tracking hypothesis) towards a solution of the video-tracking problem. Given the features extracted in the current frame and a score function that defines the quality of a candidate state, gradient-based methods iteratively refine the state estimate and converge to a local maximum of the score (Figure 5.1).

Common algorithms to solve this iterative optimisation problem are based on gradient descent [2, 3] or on expectation maximisation [4]. A convenient starting point (*initialisation*) in each frame is the previous state estimate. This naturally enforces a temporal constraint on the target motion.

5.2.1.1 *Kanade–Lucas–Tomasi (KLT) tracker*   As discussed in Section 4.3.1, a simple target appearance model is the *template* [3]. The Kanade–Lucas–Tomasi (KLT) tracker can be used to estimate the state using a template.

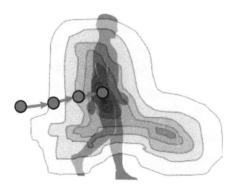

Figure 5.1 A gradient-based localisation strategy uses image information to improve the predicted state estimate.

Without loss of generality, let the target area be a square window of fixed size $N = (2\mathcal{W} - 1) \times (2\mathcal{W} - 1)$. The tracking problem can be reduced to the estimation of pure translational motion, where the state can be defined as

$$x_k = (u_k, v_k).$$

Let the coordinate system of the template, $I_T(.)$, be aligned with the coordinate system of the image $I_k, \forall k$. Given the initial candidate $\tilde{x}_k^{(0)}$ at time k as the estimated displacement x_{k-1} at time $k - 1$, the state x_k at time index k can be decomposed as [3]

$$x_k = \tilde{x}_k^{(0)} + \Delta x_k, \tag{5.1}$$

where Δx_k is a small displacement added to the previous displacement. Also, by imposing the constant-illumination constraint [3], one can consider the discrepancy in appearance between the template and the window centred around the state x_k as noise, that is

$$I_k(w) = I_T(w - x_k) + n_k(w) = I_T(w - \tilde{x}_k^{(0)} - \Delta x_k) + n_k(w), \tag{5.2}$$

with $|w - x_k|_1 < \mathcal{W}$ and $|.|_1$ denoting the L^1 norm; w is a pixel location in the image and $n_k(w)$ models additive noise over the pixel values.

Consequently, the tracking problem reduces to a search for the small displacement Δx_k that minimises the error between the image area implied by the current best estimate for x_k and the template, that is

$$\epsilon(\Delta x_k) = \sum_{|w - x_k|_1 < \mathcal{W}} \left[I_T(w - \tilde{x}_k^{(0)} - \Delta x_k) - I_k(w) \right]^2. \tag{5.3}$$

For small values of Δx_k we can approximate the template function $I_T(.)$ with its Taylor series centred around x_{k-1}, truncated to the linear term

$$I_T(w - \tilde{x}_k^{(0)} - \Delta x_k) \approx I_T(w - \tilde{x}_k^{(0)}) + b' \Delta x_k, \tag{5.4}$$

where b' is the transpose of the template gradient

$$b = \frac{\partial I_T(w - \tilde{x}_k^{(0)})}{\partial w}$$

and the template matrix is reorganised as a column vector. Then, by substituting the latter in Eq. (5.3), one can obtain

$$\epsilon(\Delta x_k) = \sum_{|w-x_k|_1 < \mathcal{W}} \left[I_T\left(w - \tilde{x}_k^{(0)}\right) - I_k(w) - b(w)'\Delta x_k \right]^2. \tag{5.5}$$

Now ϵ is a quadratic function of Δx_k and therefore Eq. (5.5) can be minimised in closed form by solving for

$$\frac{\partial \epsilon(\Delta x_k)}{\partial \Delta x_k} = 0.$$

This results in

$$\Delta x_k = \frac{\sum_{|w-x_k|_1 < \mathcal{W}} \left[I_T\left(w - \tilde{x}_k^{(0)}\right) - I_k(w) \right] b(w)}{\sum_{|w-x_k|_1 < \mathcal{W}} b(w)'b(w)}. \tag{5.6}$$

Due to the Taylor approximation, the estimate Δx_k might not correspond to a local minima of the error. It is therefore necessary to substitute $\tilde{x}_k^{(0)}$ with $\tilde{x}_k^{(0)} + \Delta x_k$ and to iterate Eq. (5.6) till convergence [3].

Note that the formulation above can be extended to more complex transformations than pure translations. When the state x represents an arbitrary linear transformation of the template region, one can compute a linear approximation like Eq. (5.4) and derive a similar iterative procedure.

To summarise, Figure 5.2 illustrates a sample tracking procedure based on KLT. Given the initial target state x_0 and the template model, for each incoming frame the KLT tracker compares the template with the image and

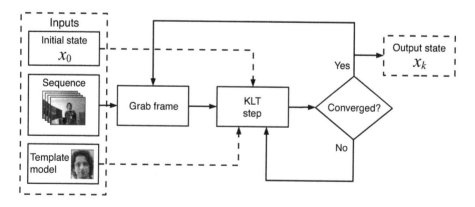

Figure 5.2   Kanade-Lucas-Tomasi (KLT) localisation procedure. The initial state estimate is refined using information from the template model and the current frame.

performs the optimisation step of Eq. (5.6). If the convergence conditions (usually based on error and step size) are not met, KLT updates the state estimate and performs another optimisation step; otherwise KLT outputs the state estimate x_k and proceeds to processing the next frame.

It is important to note that the local convergence properties of the method impose a constraint on the target motion that can be catered for. In fact, for translations that are larger than the template size, the difference between the template and the image on the denominator of Eq. (5.6) is computed over a region that does not contain the target. Thus the step estimate is uncorrelated with the real target motion. In practice the estimate tends to degrade much earlier. This depends on the method used to compute the image gradient (e.g. on the kernel size, see Section 3.3), and on the linear approximation of Eq. (5.4) becoming less and less accurate when the motion magnitude increases.

5.2.1.2 *Mean shift (MS) tracker*
When the target representation is based on colour histograms, a mean shift (MS) algorithm [2] can be used to iteratively minimise the distance in Eq. (4.5) using gradient information. Minimising Eq. (4.5) corresponds to maximising Eq. (4.6) [2]. Similarly to KLT, the MS procedure uses the previous estimate as initial state candidate $x_k^{(0)}$, that is

$$x_k^{(0)} = x_{k-1}.$$

Using Eq. (4.4) and computing the Taylor expansion of the Bhattacharyya coefficient about $r_k(x_k^{(0)})$ with respect to the centroid y_k only, one obtains

$$\rho\left[r_k(x_k), r_\mathcal{M}\right] \approx \frac{1}{2} \sum_{j=1}^{N_b} \sqrt{r_{k,j}\left(x_k^{(0)}\right) r_{\mathcal{M}j}} + \frac{C_h}{2} \sum_{i=1}^{n_h} v_i \kappa \left(\left\|\frac{y_k - w_i}{h_k}\right\|^2\right),$$

$$(5.7)$$

where

$$v_i = \sum_{j=1}^{N_b} \sqrt{\frac{r_{\mathcal{M}j}}{r_{k,j}\left(x_k^{(0)}\right)}} \delta\left[b(I_k, w_i) - j\right].$$ $$(5.8)$$

As the first term of the right-hand side in Eq. (5.7) does not depend on y_k, only the second term needs to be minimised [2]. This can be done by solving for the gradient with respect to y_k to be zero. At each iteration, the estimated target centroid shifts from $y_k^{(0)}$ to $y_k^{(1)}$, the new location, defined as

$$y_k^{(1)} = \frac{\sum_{i=1}^{n_h} w_i v_i g\left(\left\|\frac{y_k^{(0)} - w_i}{h_k^{(0)}}\right\|^2\right)}{\sum_{i=1}^{n_h} v_i g\left(\left\|\frac{y_k^{(0)} - w_i}{h_k^{(0)}}\right\|^2\right)}. \tag{5.9}$$

If

$$g(a) = -\kappa'(a),$$

then $y_k^{(1)} - y_k^{(0)}$ is in the direction of the gradient.

The iterative process continues by substituting $y_k^{(0)}$ with $y_k^{(1)}$ and stops when

$$\left\|y_k^{(1)} - y_k^{(0)}\right\| < \epsilon.$$

Usually $\epsilon = 1$ pixel. Note that the maximum area where the target can be searched for is defined by the size of the kernel κ (see Eq. 5.9). For this reason, if the shift of the target centre is larger than the kernel size, the track is likely to be lost (Figure 5.12). Note that this observation is the same we did for template-based tracking. The block diagram in Figure 5.3 summarises the localisation procedure based on MS.

5.2.1.3 Discussion   It can be seen that the difference between the MS procedure (Figure 5.3) and the KLT procedure (Figure 5.2) is the extra block that computes the colour histogram. The procedure of most trackers that are based on recursive optimisation is similar to that of MS, with the appropriate feature extraction and optimisation steps.

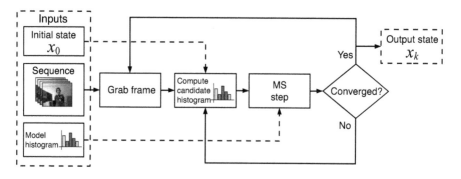

Figure 5.3   Mean Shift (MS) localisation procedure, the initial state estimate is refined using information from the histogram model and the histogram of the candidate region computed over the current frame.

In general, as well-designed optimisation algorithms converge in a small number of iterations, single-hypothesis methods are computationally inexpensive. However, as the state estimate heavily depends on the initialisation, single hypothesis localisation behaves poorly in the case of occlusions. Furthermore, when the model-candidate distance function is *multi-modal* over the area spanned by the kernel (Figure 4.6), a single-hypothesis algorithm may converge to a wrong image location. Finally, in case of convergence to an incorrect estimate, a recovery is usually unlikely, as shown in Figure 5.12(a).

5.2.2 Bayes tracking and the Kalman filter

Bayes trackers address the uncertainty problem by modelling the state x_k and the observation z_k as two stochastic processes [5].

Under the Markovian assumptions (i.e. given the observations $z_{1:k-1}$, the current state x_k depends only on its predecessor x_{k-1} and z_k depends only on the current state x_k (Figure 5.4)), the recursion is fully determined by the observation equation $\mathbf{g}_k$

$$z_k = \mathbf{g}_k(x_k, n_k), \tag{5.10}$$

and by the dynamics of x_k, defined in the state equation $\mathbf{f}_k$, as

$$x_k = \mathbf{f}_k(x_{k-1}, m_{k-1}), \tag{5.11}$$

where $\{m_k\}_{k=1,\dots}$ and $\{n_k\}_{k=1,\dots}$ are independent and identically distributed process noise sequences.

The goal of a Bayes tracker is to estimate $p_{k|k}(x_k|z_{1:k})$, the *pdf* of the object being in state x_k, given all the observations z_k up to time k [5]. The estimation is performed recursively in two steps, namely prediction and update:

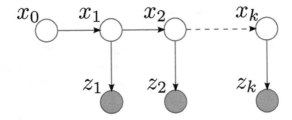

Figure 5.4 Graphical model of the tracking dependencies under the Markovian assumption. Each circle represents a random vector; the arrows show the dependencies between random vectors; filled circles represent the observables of the tracking inference [6]. The current observation z_k depends on the current state x_k only. The current state x_k depends on the previous state x_{k-1} only.

- The *prediction step* uses the dynamic model defined in Eq. (5.11) to obtain the prior *pdf* via the Chapman–Kolmogorov equation:

$$p_{k|k-1}(x_k|z_{1:k-1}) = \int f_{k|k-1}(x_k|x_{k-1})p_{k-1|k-1}(x_{k-1}|z_{1:k-1})dx_{k-1},$$

(5.12)

with $p_{k-1|k-1}(x_{k-1}|z_{1:k-1})$ known from the previous iteration and the transition density $f_{k|k-1}(x_k|x_{k-1})$ determined by Eq. (5.11) and by knowing the statistics of n_k.

- The *update step* uses Bayes' rule once the observation z_k is available, so that

$$p_{k|k}(x_k|z_{1:k}) = \frac{g_k(z_k|x_k)p_{k|k-1}(x_k|z_{1:k-1})}{\int g_k(z_k|x_k)p_{k|k-1}(x_k|z_{1:k-1})dx_k},$$

(5.13)

where $g_k(z_k|x_k)$ is determined by Eq. (5.10) and by knowing the statistics of m_{k-1}.

5.2.2.1 The Kalman filter

In general, Eq. (5.12) and Eq. (5.13) cannot be determined analytically. A well-known solution, the Kalman Filter, is available under an assumption of linearity for Eq. (5.11) and Eq. (5.10) and Gaussianity of the prior $p_{k-1|k-1}(x_{k-1}|z_{1:k-1})$ and of the two noise processes, n_k and m_{k-1} [1].

In this case, Eq. (5.10) and Eq. (5.11) rewrite as

$$z_k = G_k x_k + n_k$$

(5.14)

and

$$x_k = F_k x_{k-1} + m_{k-1},$$

(5.15)

where G_k and F_k are user-supplied matrices defining the linear relationship between consecutive states and between states and observations and the two noise processes n_k and m_{k-1} have zero mean and covariances R_k and Q_k respectively.

Given the first- and second-order statistics, and given the mean $\bar{x}_{k-1}$ and the covariance P_{k-1} of the prior $p_{k-1|k-1}(x_{k-1}|z_{1:k-1})$, from the linear relationship of Eq. (5.15) and the prediction step of Eq. (5.12), we obtain the statistics of the prediction density in the form of the mean prediction

$$\bar{x}_{k|k-1} = F_k \bar{x}_{k-1},$$

(5.16)

and the prediction covariance

$$P_{k|k-1} = F_k P_{k-1} F_k' + Q_k, \tag{5.17}$$

where the latter propagates the prior uncertainty through the state transition (Eq. 5.14) and adds to it the uncertainty given by this step (i.e. Q_k). Also, from Eq. (5.14) one can compute the predicted measurement $\hat{z}_k$ as

$$\hat{z}_k = G_k \bar{x}_{k|k-1}.$$

When the new measurement z_k becomes available, from Eq. (5.13) one can derive the mean residual

$$\bar{r}_k = z_k - \hat{z}_k,$$

the residual covariance

$$S_k = G_k P_{k|k-1} G_k' + R_k,$$

and K_k, the Kalman gain, as

$$K_k = P_{k|k-1} G_k' S_k^{-1}.$$

Finally, to complete the recursion, the first- and second-order statistics of the posterior are the mean estimate

$$\bar{x}_k = \bar{x}_{k|k-1} K_k \bar{r}_k, \tag{5.18}$$

and the posterior covariance

$$P_k = (I - K_k G_k) P_{k|k-1}. \tag{5.19}$$

In summary:

- When the assumptions hold (linearity for Eq. (5.11) and Eq. (5.10) and Gaussianity of the prior and of the two noise processes), the Kalman filter is optimal in terms of minimum mean square error on the state estimate [5].

- When the Gaussianity assumption does not hold, the Kalman filter produces a reasonable solution if the underlying distribution is well described by its first two order moments (i.e. mean and covariance).[7]

---

[7] This is usually the case when the distribution is unimodal.

- When the linearity assumptions do not hold, sub-optimal solutions are possible [5]. If $\mathbf{g}_k$ and $\mathbf{f}_k$ are differentiable, a first-order Taylor expansion can linearise the state and observation functions (i.e. $\mathbf{g}_k$ and $\mathbf{f}_k$); the result is a suboptimal filter known as the *Extended Kalman Filter (EKF)* [7].

However, the observation and state functions $\mathbf{g}_k$ and $\mathbf{f}_k$ may not be differentiable. Also, due to the Taylor approximation, the filter may quickly diverge from the optimum. Finally, as it is often the case in tracking scenarios, the posterior is not uni-modal, therefore the Kalman assumption oversimplifies the tracking problem and underestimates the amount of ambiguity in the data. To cope with these problems, multi-hypothesis localisation methods can be used, as described in the next section.

5.3 MULTIPLE-HYPOTHESIS METHODS

Multiple hypothesis localisation (MHL) methods generate multiple tracking hypotheses for each frame (Figure 5.5). These hypotheses are then validated against the image measurements and against a motion model of the object. Unlikely hypotheses are pruned from frame to frame, while the most likely state hypotheses are propagated.

Multi-hypothesis methods range from approaches that select the hypotheses based on simple heuristics to more complex solutions that propagate the hypotheses based on a probabilistic framework.

The most popular probabilistic multi-hypothesis localisation algorithm is particle filter [5], a Monte Carlo approximation of the Bayes tracking recursion. The use of multiple hypotheses allows algorithms like the particle filter to better cope with multi-modal score functions, as those generated by occlusions or by the presence of clutter (see Figure 4.6).

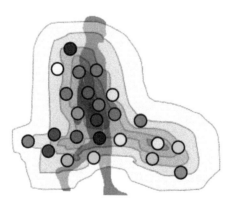

Figure 5.5 Multiple-hypothesis localisation (MHL) methods draw multiple hypotheses and then image information is used to assess the quality of each sample.

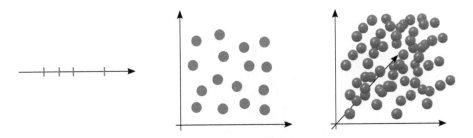

Figure 5.6 Sampling multi-dimensional state spaces. The number of hypotheses required to explore the state space grows exponentially with the number of dimensions. Left: 1D space (4 samples); centre: 2D space (16 samples); right: 3D space (64 samples).

Compared to single-hypothesis localisation methods, multiple-hypothesis localisation (MHL) methods are computationally more expensive and impose stronger limitations on the dimensionality of the state space. In fact, the number of hypotheses that are necessary to explore multi-dimensional state spaces *grows exponentially* with the number of dimensions in the state space. In the pictorial example of Figure 5.6, if we assume that all the parameters in the state space are equally important, the $4^1 = 4$ hypotheses used to sample a 1D state space become $4^2 = 16$ hypotheses for two dimensions and $4^3 = 64$ hypotheses in a 3D state space.

The major difference among the various multi-hypothesis localisation methods is the strategy used to select the tracking hypotheses, as described in the next sections.

5.3.1 Grid sampling

When the tracking task is simple and computational resources are available, a simple brute-force approach may be sufficient to solve the video-tracking problem. Grid-based localisation methods select the hypotheses in a deterministic fashion with the hypotheses positioned on a regular grid over the state space, as shown in Figure 5.7 for a 2D space.

A procedure to construct a simple recursive tracker based on grid sampling is summarised by the following steps:

1. Define a search window W centred on the previous best estimate x_{k-1}.

2. Divide the search window using a regular grid.

3. Evaluate the hypotheses of the target being present in each state defined by the grid using, for example, one of the matching scores defined in Section 4.3.

4. Select the best hypothesis as the current state x_k.

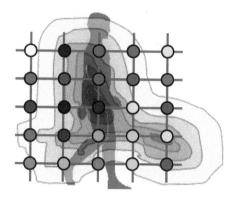

Figure 5.7 Example of multiple tracking hypotheses that are drawn using a regular sampling grid.

Note that similar procedures are common in video compression [8] when using block matching for motion estimation. To exploit the temporal redundancy between subsequent frames, first the image is split in rectangular blocks (e.g. composed of 16 × 16 pixels). Then the position of each block (i.e. its vertical and horizontal coordinates) is tracked in the previous or in the next frame. Exhaustive full-search block-matching algorithms use a sampling grid with one-pixel spacing. Next, as both the candidate regions and the template block have the same number of pixels, the evaluation of a tracking hypothesis can be done by calculating the L_1 norm between the vectors concatenating the pixel values.[8] L_1 is usually chosen as it is less computationally expensive than higher-order norms (e.g. L_2).

The precision of algorithms based on grid sampling and the selection of the fittest depends on the spacing of the grid. More accurate tracking can be achieved with a denser grid obtained by up-sampling each image frame using interpolation [9] and then applying a pixel-wide grid sampling on the larger image.

Although a full search throughout the grid is guaranteed to return the optimal result, it is practical only when the evaluation of a single candidate is inexpensive. In fact, evaluating all the hypotheses defined by the grid to find the best solution can be computationally too expensive. When computational resources are not sufficient, suboptimal solutions exist. For example, a popular approach is hierarchical sampling, where first a coarser grid of samples is evaluated and then recursively finer grids of hypotheses centred on the current best hypothesis are evaluated. An example of a three-stage hierarchical sampling is given in Figure 5.8. In this figure, the red dot indicates the best hypothesis at each step.

---

[8] Note that this is equivalent to taking the pixel information in the block as a template model (see Section 4.3.1).

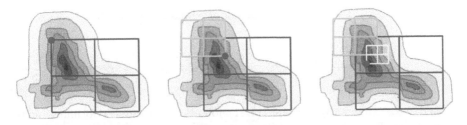

Figure 5.8 Hierarchical grid sampling procedure. From left to right: a finer and finer grid is used to refine the position of the current best hypothesis (i.e. the red dot).

5.3.2 Particle filter

When $\mathbf{f}_k$ and $\mathbf{g}_k$ (Eq. 5.11 and Eq. 5.10) are non-linear, time-varying functions, we can consider a solution based on a Monte Carlo integration [5]. An application of this framework to visual target tracking first appeared in [10].

The densities $p_{k|k}(x_k|z_{1:k})$ are approximated with a sum of L_k Dirac δ functions (the particles) centred in

$$\left\{ x_k^{(i)} \right\}_{i=1}^{L_k}$$

as

$$p_{k|k}(x_k|z_{1:k}) \approx \sum_{i=1}^{L_k} \omega_k^{(i)} \delta\left(x_k - x_k^{(i)} \right), \tag{5.20}$$

where

$$\left\{ \omega_k^{(i)} \right\}_{i=1}^{L_k}$$

are the weights associated with the particles and are defined as

$$\omega_k^{(i)} \propto \frac{p_{k|k}(x_k^{(i)}|z_{1:k})}{q_k(x_k^{(i)}|z_{1:k})} \quad i = 1, \ldots, L_k. \tag{5.21}$$

$q_k(.)$ is the importance density function defined as the density that generated the current set of particles.

Let us assume that $p_{k-1|k-1}(x_{k-1}|z_{1:k-1})$ is approximated by the set of particles and associated weights

$$\left\{ \omega_{k-1}^{(i)}, x_{k-1}^{(i)} \right\}_{i=1}^{L_{k-1}},$$

as in Eq. (5.20). By substituting this approximation in Eq. (5.12) we obtain

$$p_{k|k-1}(x_k|z_{1:k-1}) \approx \sum_{j=1}^{L_{k-1}} \omega_{k-1}^{(j)} f_{k|k-1}(x_k|x_{k-1}^{(j)}). \tag{5.22}$$

Then from Eq. (5.13) and Eq. (5.21) follows the recursive formulation to propagate the particles and their weights [10] as

$$\omega_k^{(i)} \propto \frac{g_k(z_k|x_k^{(i)}) \sum_{j=1}^{L_{k-1}} \omega_{k-1}^{(j)} f_{k|k-1}(x_k^{(i)}|x_{k-1}^{(j)})}{q_k(x_k^{(i)}|z_{1:k})}. \tag{5.23}$$

In the *CONDENSATION* algorithm from Isard and Blake [10] the particles are drawn from the predicted prior, i.e.

$$q_k(x_k|z_{1:k}) = p_{k|k-1}(x_k|z_{1:k-1}). \tag{5.24}$$

This reduces the weighting equation (Eq. 5.23) to

$$\omega_k^{(i)} \propto g_k(z_k|x_k^{(i)}), \tag{5.25}$$

where the weights are now proportional to the likelihood.

An alternative formulation [5] of the weight update is possible by applying a Monte Carlo approximation to the joint posterior

$$p_{k|k}(x_{1:k}|z_{1:k}) \approx \sum_{i=1}^{L_k} \omega_k^{(i)} \delta\left(x_{1:k} - x_{1:k}^{(i)}\right), \tag{5.26}$$

where the weights are defined as

$$\omega_k^{(i)} \propto \frac{p_{k|k}(x_{1:k}^{(i)}|z_{1:k})}{q_k(x_{1:k}^{(i)}|z_{1:k})} \quad i = 1, \ldots, L_k. \tag{5.27}$$

Then, under Markovian assumptions, by decomposing the importance sampling function $q_k(x_{1:k}|z_{1:k})$ as

$$q_k(x_{1:k}|z_{1:k}) = q_k(x_k|x_{1:k-1}, z_{1:k}) q_k(x_{1:k-1}|z_{1:k-1}) \tag{5.28}$$

and assuming

$$q_k(x_k|x_{1:k-1}, z_{1:k}) = q_k(x_k|x_{k-1}, z_k),$$

it follows that

$$\omega_k^{(i)} \propto \frac{\omega_{k-1}^{(i)}}{a_{k-1}^{(i)}} \frac{g_k(z_k|x_k^{(i)}) f_{k|k-1}(x_k^{(i)}|x_{k-1}^{(i)})}{q_k(x_k^{(i)}|x_{k-1}^{(i)}, z_k)}. \tag{5.29}$$

To discard particles with lower weights, a resampling step can be applied before propagation. The resampling step draws the

$$\left\{ x_k^{(i)} \right\}_{i=1}^{L_k}$$

from the set

$$\left\{ x_{k-1}^{(i)} \right\}_{i=1}^{L_{k-1}}$$

according to the resampling function [11]

$$\left\{ a_{k-1}^{(i)} \right\}_{i=1}^{L_{k-1}}.$$

The *resampling function* defines the probability of each particle $x_{k-1}^{(i)}$ to generate a new sample at time k. The weighted estimate of the filtering posterior $p_{k|k}(x_k|z_{1:k})$ is obtained by marginalising $x_{1:k-1}$ out of Eq. (5.26).

The weights $\omega_k^{(i)}$ are constant in the marginalisation as they do not depend on $x_{1:k-1}$, but on the particle positions $x_k^{(i)}$ and $x_{k-1}^{(i)}$ only. Consequently the approximated filtered posterior of Eq. (5.20) inherits the same weights as the joint posterior of Eq. (5.26).

The two *recursive update equations* (Eq. 5.23 and Eq. 5.29) differ mainly in the importance sampling function. By accounting for the dependency of q_k with the previous state x_{k-1} the update of the weight $\omega_k^{(i)}$ becomes dependent only on $x_{k-1}^{(i)}$ and not on the global sample set, as in Eq. (5.23). Although the computation of Eq. (5.29) requires less multiplications than Eq. (5.23), computing q_k requires extra care to account for dependencies between the factorised importance sampling function and the previous time steps.

The filters, based on Monte Carlo sampling and recursive Bayes equations, are known as *particle filters (PFs)*. The estimate of the state is usually computed either by taking the maximum a posteriori estimate, i.e. the particle with largest weight, or by computing the expectation over the weighted particles as

$$\mathbf{E}[x_k|z_{1:k}] \approx \frac{1}{L_k} \sum_{i=1}^{L_k} \omega_k^{(i)} x_k^{(i)}. \tag{5.30}$$

Figure 5.9 shows an example of a particle filter with resampling. Resampling is an optional step that is applied to prune unlikely hypotheses

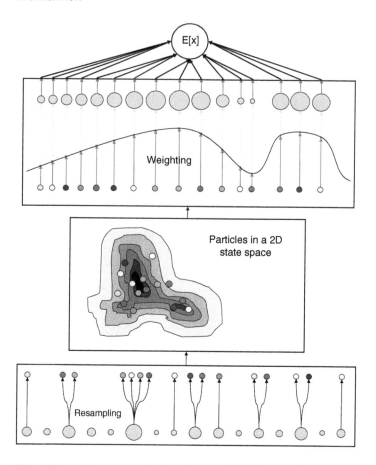

Figure 5.9 Schematic representation of a particle filter with systematic resampling. From bottom to top: the filter resamples the particles and shifts them in the state space according to the importance sampling function ($q_k(x_k|z_{1:k})$ or $q_k(x_k|x_{k-1}, z_k)$); then, using the new observation, the filter updates the weights according to Eq. (5.23) or Eq. (5.29). IEEE © [12].

from the particle set and to avoid that after a few iterations all but one particle have negligible weights. This phenomenon is usually known as *particle impoverishment* [5].

Unlike the Kalman filter, PF [5, 13, 14] can deal with multi-modal *pdf*s and thus can recover from short-term occlusions. However, the tracking result depends heavily on the chosen parameter setting, which in turns depends on the scene content. Moreover, the number of particles required to model the underlying *pdf* increases exponentially with the dimensionality of the state space, thus dramatically increasing the computational loads. The efficiency of PF depends in fact on the distribution of the samples in the state space (i.e. on q_k).

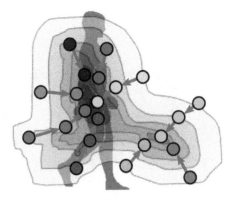

Figure 5.10 Hybrid localisation algorithms use image information and gradient-like optimisation to improve the initial set of hypotheses.

5.3.3 Hybrid methods

To reduce the computational complexity while improving the sampling of multiple-hypothesis methods, hybrid localisation (HL) algorithms exist that combine MHL with single-hypothesis localisation (SHL) strategies [12, 15–17]. The hypotheses generated by MHL are refined by a local optimisation step that uses information from the current frame (Figure 5.10). The multiplicity of the first set of hypotheses helps in coping with occlusions and clutter, while the refining optimisation step improves the quality of each hypothesis and consequently the computational efficiency of the localisation step.

When using a multiple-hypothesis tracker like PF, particles positioned far from the target area might bias the state estimator. As anticipated in Section 5.3.2, a common solution adopted by the CONDENSATION [5,10] algorithm (here referred to as particle filter, CONDENSATION implementation (PF-C)) is to redraw the particles from the prior (see Eq. 5.24). However, sampling in PF-C does not account for information from the most recent measurement. As a consequence, when the dynamic model is not accurate, the area of the state space around the target is not densely sampled.

5.3.3.1 *Sampling strategies* To account for the latest measurement, several sampling strategies have been proposed [17–20], based, for example, on Markov Chain Monte Carlo (MCMC), simulated annealing and the EKF:

- MCMC samplers can improve sampling efficiency with high-dimensional state spaces [17,21,22]. Choo and Fleet [17] showed that with 10 or more degrees of freedom, MCMC is more efficient than PF. However, due to the relatively large number of steps necessary for MCMC to converge, no improvements in terms of efficiency are reported on low-dimensional state spaces [17].

- In simulated annealing [15], first the particles are randomly spread over the state space, then a layered procedure redraws the samples whose number is proportional to their likelihood.

- When the relationship between state and measurement can be linearised, an alternative is to sample from the Gaussian estimate computed by an EKF associated with each particle [18]. EKF can be substituted with an unscented transform that does not require linearisation [18,23]. Both methods assume that the modes of the *pdf* are well represented by their first- and second-order moments.

A different approach to particle sampling is to derive the particles according to point estimates of the gradient of either the posterior or the likelihood [17, 19, 21, 22, 24, 25]. When the appearance model is a template, optical flow can be used to drive the particles towards peaks of likelihood. However, as motion blur can affect the accuracy of optical flow, the particle-shifting procedure is enabled only when the momentum of the object is small [19].

A more principled solution, known as the kernel particle filter, uses kernel density estimation to produce from the particle set a continuous approximation of the posterior *pdf*. Then, the sample-based mean shift (MS), a kernel-based iterative procedure, is used to approximate the gradient of the *pdf* and to climb its modes [20]. However, as the accuracy of the density estimate and of its gradient depends on the sampling rate, a reduction in the number of samples may affect the quality of the final approximation. Another solution is to use colour MS to drive the particles to improve sampling efficiency with low-dimensional state spaces and highly manoeuvrable targets [25]. This is the solution detailed below.

5.3.3.2 Hybrid-particle-filter-mean-shift (HY) tracker

Let the appearance model be the normalised colour histogram,

$$r_{\mathcal{M}} = r_0(x_0)$$

defined in Eq. (4.4) and let the weight update procedure of PF be the one defined in Eq. (5.23).

To select the particle states, the hybrid-particle-filter-mean-shift (HY) tracker (Figure 5.11) first draws a set of samples

$$\left\{ \tilde{x}_k^{(i)} \right\}_{i=1}^{L_k}$$

from the approximated predicted *pdf* $p_{k|k-1}(x_k|z_{1:k-1})$ (see Eq. 5.22), using for example a zero-order Gaussian state transition model

$$x_k = x_{k-1} + m_{k-1}, \tag{5.31}$$

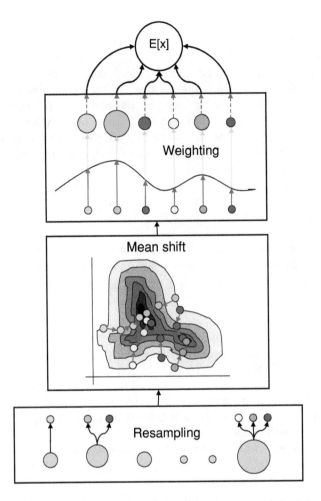

Figure 5.11 Schematic representation of the hybrid-particle-filter-mean-shift (HY) tracker. The particles are first drawn from the prior, then their position is refined by the MS optimisation procedure. IEEE © [12].

where m_{k-1} is a multi-variate Gaussian random variable with 0 mean vector and diagonal covariance Σ.

We can also express this by saying that the state transition probability

$$f_{k|k-1}(x_k|x_{k-1}) = \mathcal{N}(x_k; x_{k-1}, \Sigma)$$

is Gaussian distributed and $\mathcal{N}$ denotes a Gaussian evaluated in x_k with mean x_{k-1} and covariance Σ.

The choice of a relatively uninformative motion model is usually motivated by the necessity to track highly manoeuvring targets. On the one hand, higher-order models are desirable when the motion can be predicted with good accuracy; however, if a wrong prediction is made, most particles will be distributed in regions of low interest, thus compromising the robustness of the algorithm. On the other hand, a zero-order model needs a larger number of particles as the search area is not constrained by a strong prediction.

In both cases, the sampling based on the predicted *pdf* $p_{k|k-1}(x_k|z_{1:k-1})$ only (CONDENSATION) is blind to the current observation z_k. A sampling criterion that concentrates the particles in regions with high likelihood is expected to increase their efficiency. To this end, a method based on the MS iterative search can be used [25].

After propagation, each particle in

$$\left\{ \tilde{x}_t^{(i)} \right\}_{i=1}^{L_k}$$

is independently re-located in the position state subspace using MS. The traditional MS [2] iteratively minimises the distance in Eq. (4.5) using gradient information. The algorithm is initialised at

$$x^{(0)} = \tilde{x}_k^{(i)},$$

the particle position, and converges to the nearest local minimum.

The maximum area where the target can be searched is the kernel size (Eq. 5.9). For this reason, if the shift of the target centre is larger than the kernel size, the track is likely to be lost. The multiple MS initialisation produced by PF sampling overcomes this problem. Therefore the hyper-volume of the state space under analysis is defined by the state equation (Eq. 5.31) and enhanced by MS iterations.

To summarise, let MS be defined as the operator

$$\mathcal{M} : R^d \rightarrow R^d,$$

where d is the state space dimensionality. The final set of particles

$$\left\{ x_k^{(i)} \right\}_{i=1}^{L_k}$$

is obtained by applying

$$x_k^{(i)} = \mathcal{M}(\tilde{x}_k^{(i)}). \tag{5.32}$$

$\mathcal{M}$ takes as input each particle $\tilde{x}_k^{(i)}$ and modifies the state position y by guiding each particle over the position subspace independently from all others.

To weight the particles, the likelihood is computed using the colour histograms model, and the distances from the model defined in Eq. (4.5) and already computed in the MS iterations as

$$g_k(z_k|x) = e^{-\left(\frac{d[r_k(x),r_{\mathcal{M}}]}{\sigma} \right)^2}. \tag{5.33}$$

The value of σ depends on the histogram dimensionality. The higher the dimensionality, the larger the average distance of Eq. (4.5), hence the higher the value for σ used to obtain a smoother likelihood.

In *CONDENSATION* [5], here referred to as PF-C,

$$q_k(x_k|z_{1:k}) = p_{k|k-1}(x_k|z_{1:k-1})$$

is the predicted prior and the weighting reduces to Eq. (5.25) where the weights are proportional to the likelihood.

Note that in the hybrid approach, as the particles are shifted by the MS procedure, the importance sampling function $q_k(.)$ is no longer the predicted prior. In fact, the hybrid tracker can be interpreted as a particle filter with a *semi-deterministic importance density* $q_k(x_k^{(i)}|z_{1:k})$ (see Eq. 5.21).

Weighting according to Eq. (5.25) would introduce a bias in the posterior approximation. To prevent this, we approximate, as in kernel particle filter [20],

$$q_k(x_k|z_{1:k}) \approx \hat{q}_k(x_k),$$

using a Gaussian kernel density estimation as

$$\hat{q}_k(x_k) = \frac{1}{N_s} \sum_{i=1}^{N_s} \hat{q}_{k\,N_s,i}(x_k), \tag{5.34}$$

where

$$\hat{q}_{k\,N_s,i}(x_k) = \left(\left(h_{N_s} \sqrt{2\pi} \right)^d \sqrt{\det(\hat{\Sigma})} \right)^{-1} \times$$

$$\times \exp\left[\frac{-1}{2h_{N_s}^2} \left(x_k - x_k^{(i)} \right)' \hat{\Sigma}^{-1} \left(x_k - x_k^{(i)} \right) \right]. \tag{5.35}$$

$\hat{\Sigma}$ is the covariance matrix calculated over

$$\left\{ x_k^{(i)} \right\}_{i=1}^{N_s}.$$

The kernel bandwidth h_n is the one giving the optimal rate of convergence in probability to zero of the integrated squared error [26]

$$\int \left(\hat{q}_k(x_k) - q_k(x_k) \right) dx_k.$$

Hence

$$h_{N_s} = c \cdot N_s^{\frac{-1}{(d+4)}}, \tag{5.36}$$

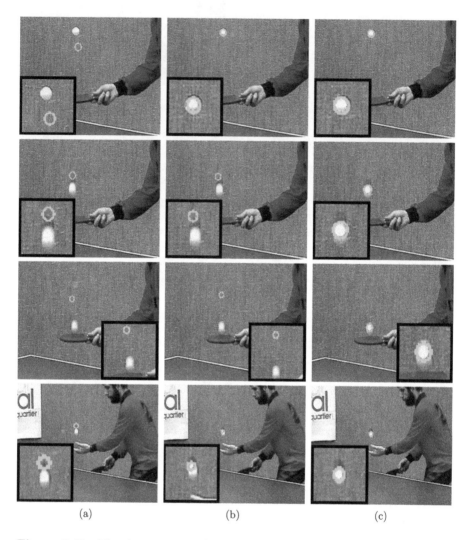

Figure 5.12   Visual comparison of tracking performance; (a) mean shift (MS); (b) particle filter-CONDENSATION (PF-C); (c) hybrid tracker (HY). Unlike MS and PF-C, HY successfully tracks the target under abrupt accelerations. Reproduced with permission of Elsevier [25].

where the constant c is

$$c = \left(\frac{4}{d+2} \right)^{1/(d+4)}.$$

Finally, the best state at time k is estimated using the discrete approximation of Eq. (5.20). The most common solution is the Monte Carlo approximation of the expectation, as in Eq. (5.30), which, as opposed to PF-C, operates on particles that are concentrated near local maxima of the likelihood. This makes the particles drawn by the HY tracker more efficient than those in PF-C. The particles are filtered and selected depending on their likelihood. This makes HY more than a simple multiple initialisation of MS. In fact, HY inherits from PF the possibility to treat multi-modal *pdf*s and to recover from occlusions.

Unlike the estimation of the kernel particle filter [20], which can be critical when the number of particles is small, this estimation of the gradient is not based on the relative position of the particles, but on the colour histogram itself. Although more specific to the appearance-based tracking problem, this solution is less dependent on the density of the sampling.

To conclude, sample comparative results between MS, PF-C and HY are shown in Figure 5.12 on a target with fast accelerations. Due to these accelerations, only HY follows the target, due to the single-hypothesis refinement of the particles that allows the algorithm to adapt to *varying* target motion conditions. MS cannot track fast-moving targets with motion larger than the basin of attraction of the gradient. PF-C would need to update the state transition model to follow the varying motion of the target. For further results and discussion see Section A.2 in the Appendix.

5.4 SUMMARY

In this chapter we discussed localisation strategies and we classified them into single- and multiple-hypothesis approaches. In particular, we focused on grid-based, gradient-based and Bayesian localisation strategies. As example of Bayesian strategies, we covered the Kalman filter and the particle filter. Finally, we presented hybrid localisation strategies that use information from the current observation to improve sampling efficiency.

In the next chapter, we review approaches that fuse the contributions of multiple features to improve video-tracking results.

REFERENCES

1. R.E. Kalman. A new approach to linear filtering and prediction problems. *Journal of Basic Engineering*, 82(1):35–45, 1960.

2. D. Comaniciu, V. Ramesh and P. Meer. Kernel-based object tracking. *IEEE Transactions on Pattern Analysis and Machine Intelligence*, 25(5):564–577, 2003.

3. B.D. Lucas and T. Kanade. An iterative image registration technique with an application to stereo vision. In *Proceedings of the 7th International Joint Conference on Artificial Intelligence (IJCAI '81)*, Vancouver, BC, Canada, 1981, 674–679.

4. A.D. Jepson, D.J. Fleet and T. El-Maraghi. Robust online appearance models for visual tracking. *IEEE Transactions on Pattern Analysis and Machine Intelligence*, 25(10):1296–1311, October 2003.

5. M.S. Arulampalam, S. Maskell, N. Gordon and T. Clapp. A tutorial on particle filters for online non-linear/non-Gaussian Bayesian tracking. *IEEE Transactions on Signal Processing*, 50(2):174–188, 2002.

6. C.M. Bishop. *Pattern Recognition and Machine Learning (Information Science and Statistics)*. Berlin, Springer, 2006.

7. G. Welch and G. Bishop. An introduction to the Kalman filter. Technical report, University of North Carolina at Chapel Hill, Chapel Hill, NC, 1995.

8. I.E. Richardson. *H.264 and MPEG-4 Video Compression: Video Coding for Next Generation Multimedia* (1st Edition). New York, John Wiley & Sons, Inc., 2003.

9. R.C. Gonzalez and R.E. Woods. *Digital Image Processing* (3rd Edition). Upper Saddle River, NJ, Prentice-Hall, Inc., 2006.

10. M. Isard and A. Blake. CONDENSATION – conditional density propagation for visual tracking. *International Journal of Computer Vision*, 29(1):5–28, 1998.

11. J. Liu, R. Chen, and T. Logvinenko. A theoretical framework for sequential importance sampling and resampling. Technical report, Stanford University, Department of Statistics, 2000.

12. E. Maggio and A. Cavallaro. Hybrid particle filter and mean shift tracker with adaptive transition model. In *Proceedings of the IEEE International Conference on Acoustics, Speech, and Signal Processing*, Vol. 2, Philadelphia, PA, 2005, 221–224.

13. K. Nummiaro, E. Koller-Meier and L. Van Gool. A color-based particle filter. In *Proceedings of the Workshop on Generative-Model-Based Vision*, Copenhagen, Denmark, 2002, 53–60.

14. S. Zhou, R. Chellappa and B. Moghaddam. Appearance tracking using adaptive models in a particle filter. In *Proceedings of the Asian Conference on Computer Vision*, Melbourne, AU, 2002.

15. J. Deutscher, A. Blake and I. Reid. Articulated body motion capture by annealed particle filtering. In *Proceedings of the IEEE Conference on Computer Vision and Pattern Recognition*, Vol. 2, Hilton Head, SC, 2000, 126–133.

16. Z. Khan, T. Balch and F. Dellaert. An MCMC-based particle filter for tracking multiple interacting targets. In *Proceedings of the European Conference on Computer Vision*, Prague, Czech Republic, 2004, 279–290.

17. K. Choo and D.J. Fleet. People tracking using hybrid Monte Carlo filtering. In *Proceedings of the International Conference on Computer Vision*, Vancouver, Canada, 2001, 321–328.

18. R. van der Merwe, A. Doucet, N. de Freitas and E. Wan. The unscented particle filter. Technical Report CUED/F-INFENG/TR380, Cambridge University, Engineering Department, August 2000.

19. J. Sullivan and J. Rittscher. Guiding random particles by deterministic search. In *Proceedings of the International Conference on Computer Vision*, Vancouver, Canada, 2001, 323–330.

20. C. Chang and R. Ansari. Kernel particle filter: iterative sampling for efficient visual tracking. In *Proceedings of the IEEE International Conference on Image Processing*, Vol. 3, Barcellona, Spain, 2003, III–977–80.

21. C. Sminchisescu and B. Triggs. Covariance scaled sampling for monocular 3D body tracking. In *Proceedings of the IEEE Conference on Computer Vision and Pattern Recognition*, Vol. 1, Kauai, HI, 2001, 447–454.

22. C. Sminchisescu and B. Triggs. Hyperdynamics importance sampling. In *Proceedings of the European Conference on Computer Vision*, Vol. 1, Copenhagen, Denmark, 2002, 769–783.

23. P. Li, T. Zhang and A.E.C. Pece. Visual contour tracking based on particle filters. *Image and Vision Computing*, 21(1):111–123, 2003.

24. C. Shan, Y. Wei, T. Tan and F. Ojardias. Real time hand tracking by combining particle filtering and mean shift. In *Proceedings of the Sixth IEEE International Conference on Automatic Face and Gesture Recognition*, Killington, VT, 2004, 669–674.

25. E. Maggio and A. Cavallaro. Accurate appearance-based bayesian tracking for maneuvering targets. *Computer Vision and Image Understanding*, 113(4):544–555, 2009.

26. B.W. Silverman. *Density Estimation for Statistics and Data Analysis*, Vol. 37-1. Boca Raton, FL, CRC Press, 1988.

6

FUSION

6.1 INTRODUCTION

Multiple and possibly independent sources of information are commonly used in signal processing to improve the results of an algorithm. In particular, the exploitation of multiple visual features can help improve the accuracy and the robustness of a video tracker.

When using multiple features, an important challenge is to determine:

- their type and number
- their relative importance
- the information fusion mechanism.

Ideally, the importance of each feature should be adapted to the changes in target pose and the surrounding background. This adaptation could improve the performance under changes that are not modelled by the tracker itself and hence it would remove the need for retuning the algorithm through human interaction.

Video Tracking: Theory and Practice. Emilio Maggio and Andrea Cavallaro
© 2011 John Wiley & Sons, Ltd

In the first part of this chapter we present different strategies for fusion in video tracking. In the second part we present a general procedure for the adaptive combination of multiple representations in a particle filter framework. In this context, we describe a strategy for estimating the importance of each feature.

6.2 FUSION STRATEGIES

Multi-feature fusion in video tracking can be performed either at the tracker level or at the measurement level. While *tracker-level* fusion enables the use of a range of different trackers, fusion at the *measurement-level* avoids running multiple single-feature trackers, thus reducing the problem of having to merge possibly inconsistent or redundant tracking hypotheses.

6.2.1 Tracker-level fusion

Fusion at *tracker level* models single-feature tracking algorithms as black boxes. The video-tracking problem is redefined by modelling the interaction between outputs of black boxes, which can run in parallel or in cascade (sequentially) (Figure 6.1).

For example, multiple independent CONDENSATION algorithms can be used on each feature of the target representation, followed by the integration of the target estimates by multiplying the posterior probabilities [1] (Figure 6.1(a)). If each feature spans a separate subspace of the target state, a similar framework can also account for conditional feature dependencies [2].

The outputs of independent algorithms tracking localised parts of the target can be used as the observables of a Markov network [3], with a state variable associated with each part. Assuming linear and Gaussian interactions, the network models the compatibility between the states (i.e. the position of the parts). An algebraic criterion assesses the inter-part consistency, thus allowing the removal of inconsistent measurements.

An alternative is to perform the fusion *sequentially*, considering the features as if they were available at subsequent time instants (Figure 6.1(b)). The results of a blob detector, a colour-based mean shift (MS) tracker and a feature-point tracker can be incrementally incorporated by extended Kalman filtering [4]. The frame-by-frame measurement noise used by the filter for each feature acts as a feature reliability estimator. The measurement noise can also be estimated in a training phase [5], thus avoiding having to adapt the feature contribution over time, but reducing the flexibility of the tracker under changing scene conditions.

A summary of multi-feature tracking algorithms with fusion at the tracker level is given in Table 6.1. The table highlights the features and the fusion mechanism used by the trackers.

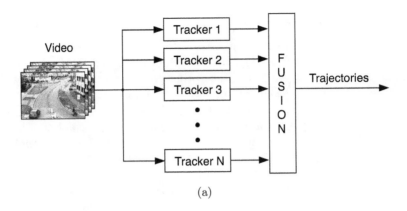

(a)

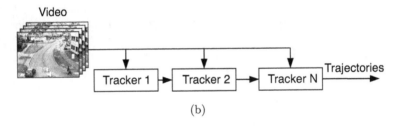

(b)

Figure 6.1 Tracker-level data fusion strategies for video tracking: (a) fusion of independent estimates from parallel trackers; (b) sequential integration of tracking estimates in a cascade architecture.

Table 6.1   Comparison of feature types and fusion strategies used in video tracking algorithms that combine multiple features at the tracker level.

Ref.	Algorithm	Features	Fusion
[1]	CONDENSATION, Kalman filter	Template, blob, colour	Non-adaptive product
[2]	Particle filter	Colour, contour	Product of *pdfs*
[3]	Kanade–Lucas, particle filter	Template	Bayesian network
[4]	Extended Kalman filter	Blob, colour, geometry	Sequential integration
[5]	CONDENSATION	Template, colour	Covariance estimation

6.2.2 Measurement-level fusion

When fusing multiple features at *measurement level*, the measurements are combined internally by the tracking algorithm. Fusion can take place with a variety of mechanisms, such as via a voting procedure, using saliency maps, Bayesian networks or the estimation of mutual information.

For example, the phase coefficients of the wavelet decomposition can be considered as multiple features forming a time-evolving template [6]. Each phase coefficient is modelled independently by a mixture of three components: (i) a stable component, (ii) a fast varying component and (iii) a component that models outliers. The fusion is performed by a *procedure* that gives more importance to stable coefficients. It is important to notice that, as all the measurements are generated with the same technique, they also present the same failure modalities [6].

As an alternative, a Markov model can be used to eliminate the feature measurements generated by clutter and to replace occluded measurements [7]. Also a *Bayesian network* can model the dependency of multiple reliability scores to evaluate different features [8]. This method requires a training phase to learn the parameters of the network.

Saliency maps of multiple features can be adaptively integrated as a *weighted average*, where the weights depend on the correlation between the saliency of each feature and the overall result [9]. The decision is based on the evaluation of the different descriptors on the whole frame and therefore this solution is applicable to single-target tracking only. Moreover, the results are valid only when consensus between the individual features is predominant [10].

To overcome this limitation, a particle-filter framework can be used, which evaluates the features only on the tracking hypotheses (particles) propagated by the algorithm. For example, multiple multi-modal features can be fused non-adaptively in a particle filter assuming conditional independence of the features, given the state [11]. In this approach, the feature contribution is held constant and the adaptivity relies on the resampling step that discards particles with low likelihood. If one can assume inter-feature independence, the contribution of each feature can be taken into account by *multiplying likelihoods* and by selecting weights based on the distance between the tracking result of each feature and the global tracking result. Each weight is used as the exponent of the corresponding likelihood [12]. Note that this solution is equivalent to creating a weighted log-likelihood mixture [13]. A similar reliability measure is used in a *voting* framework to fuse five features for visual servoing [14].

To account for cooperative feature interaction, *mutual information* can be used to quantify inter-feature agreement [15] and to assess feature reliability [16]. Feature interaction can be learned using a *graphical model* approximated by variational inference and Monte Carlo sampling [17]. Using colour and shape information, the colour state is iteratively updated by sampling the shape prior, whereas the shape state is updated by sampling the colour prior. A graphical model coupled with inter-feature belief propagation can also be used to integrate colour, shape and intensity [18], where the final output is a set of three different states (each one associated with a feature). In such a case the fusion problem is delayed to a subsequent step.

A comparison of features and fusion mechanisms at the measurement-level used in multi-feature tracking is given in Table 6.2.

6.3 FEATURE FUSION IN A PARTICLE FILTER

6.3.1 Fusion of likelihoods

The particle filtering [21] tracking approach described in Section 5.3.2 can account for multiple features N_f at the likelihood level. Suppose that we can evaluate a likelihood $g_{k,j}(z_k|x_k)$ at time k for each feature j ($j = 1, \ldots, N_f$). A solution is to compute the overall likelihood as a linear combination of the single feature likelihoods [10] as

$$g_k(z_k|x_k) = \sum_{j=1}^{N_f} \alpha_{k,j}\, g_{k,j}(z_k|x_k), \qquad (6.1)$$

where $\alpha_{k,j}$ is a mixture coefficient and

$$\sum_{j=1}^{N_f} \alpha_{k,j} = 1. \qquad (6.2)$$

The sum rule was demonstrated in the case of object classification to outperform the product rule and other classifier combinations schemes by being less sensitive to ambiguous and inconsistent measurements [22]. Similarly, in object tracking the sum rule can be a better choice as it is less sensitive to clutter and targets with similar appearance. Moreover, this strategy is in line with the assumption that humans perceive visual content through a sum of multiple features weighted by their reliability [23].

As an example, to implement the filter we describe the target area with two features (i.e. $N_f = 2$) based on histograms. The first feature is the multi-part colour histogram defined in Section 4.3.2. Colour information is complemented by information on object shape and internal edges encoded in the orientation histogram also defined in Section 4.3.2. Let us calculate the likelihood of each feature using the distance from the model histograms, defined in Eq. (4.5) as

$$g_{k,j}(z_k|x_k) = e^{-d(r_{k,j}(x_k),\, r_{\mathcal{M},j})^2/(2\sigma_j^2)}. \qquad (6.3)$$

Table 6.2   Comparison of feature types and fusion strategies used in video-tracking algorithms that combine multiple features at the measurement level.

Ref.	Algorithm	Features	Fusion
[19]	Full search on motion predicted region	Colour histogram, edge map	Non-adaptive linear combination
[20]	Trust region search	Colour histogram, edge density	Non-adaptive linear combination
[6]	EM on affine parameters	Phase of wavelet coefficients	Enhancing stable coefficients
[7]	Monte Carlo	Edge feature points on the contour	Clutter and occlusion Markov modelling
[9]	Saliency map fusion (full search)	Motion, colour, position, shape, contrast	Adaptive democratic integration
[11]	Particle filter	Colour, motion, sound	Non-adaptive likelihood factorisation
[10]	Multi-target clustered particle filter	Motion, colour, Kalman prediction, shape, contrast	Non-adaptive linear combination
[12]	Particle filter	Colour, shape	Adaptive log-likelihood mixture
[14]	Optimised search on window	Edge, disparity, colour, template, motion	Adaptive voting
[8]	Kalman filter	Colour, motion, blob	Bayesian network
[15]	Full search	Shape, colour, template	Inter-feature mutual information
[16]	Multiple hypotheses	Intensity, texture, colour	Intra-feature mutual information
[17]	Monte Carlo	Colour, shape	Co-inference learning
[18]	Monte Carlo	Colour, shape, intensity change	Inter-feature belief propagation

Note that the exponent is used to obtain a smooth likelihood, thus facilitating the final state estimation. The value of σ_j, which models the noise in the measurements, is determined experimentally. In this specific example, it is based on the fact that gradient orientation is more affected by noise than colour and that the finer the quantisation, the higher the impact of the noise.

Starting from the weight-update formulation of Eq. (5.29) we further simplify the problem by taking q_k, the importance distribution used to sample the particles, proportional to the state transition model, i.e.

$$q_k(x_k|x_{k-1}, z_k) \propto f_{k|k-1}(x_k|x_{k-1}).$$

Thus for arbitrary resampling functions

$$\omega_k^{(i)} \propto \frac{\omega_{k-1}^{(i)}}{a_{k-1}^{(i)}} g_k\left(z_k|x_k^{(i)}\right). \tag{6.4}$$

6.3.2 Multi-feature resampling

When systematic multinomial resampling is used (i.e. the particles are resampled proportionally to their weight),

$$a_{k-1}^{(i)} = \omega_{k-1}^{(i)},$$

hence

$$\omega_k^{(i)} \propto g_k\left(z_k|x_k^{(i)}\right), \tag{6.5}$$

that is, the weights are proportional to the likelihood of the observation vector [24]. This means that, in the multi-feature case, particles are drawn proportionally to the mixed likelihood weights of Eq. (6.1). When the algorithm degenerates (i.e. all but one feature give negligible contribution), most particles are resampled from a single feature, thus ignoring the other components of the mixed likelihood.

As the evaluation of the *reliability* of each feature requires a set of particles that accurately represents all the components of the mixture as defined in Eq. (6.1), it is appropriate to introduce a multi-feature resampling strategy. The resampling function is defined as

$$a_k^{(i)} = \sum_{j=1}^{N_f} \beta_{k,j}\, g_{k,j}\left(z_k|x_k^{(i)}\right) \quad i = 1, ..., L_k, \tag{6.6}$$

where

$$\beta_{k,j} = \begin{cases} \alpha_{k,j} & \text{if } \alpha_{k,j} > V/N_f \\ V/N_f & \text{otherwise} \end{cases} \quad j = 1, ... N_f. \tag{6.7}$$

The threshold $V > 0$ prevents all the particles being drawn from the likelihood distribution of a single feature if the weights become unbalanced. [9]
After thresholding, we normalise the weights so that

$$\sum_{j=1}^{N_f} \beta_{k,j} = 1. \tag{6.8}$$

Algorithm 6.1 Multi-feature adaptive particle filter

$$\left[\left\{x_{k-1}^{(i)}, \omega_{k-1}^{(i)}\right\}_{i=1}^{L_k}, \left\{\alpha_{k-1,j}\right\}_{j=1}^{N_f}\right] \rightarrow \left[\left\{x_k^{(i)}, \omega_k^{(i)}\right\}_{i=1}^{L_k}, \left\{\alpha_{k,j}\right\}_{j=1}^{N_f}\right]$$

1: Compute $\left\{a_{k-1}^{(i)}\right\}_{i=1}^{L_k}$ according to Eq. (6.6)

2: Resample the particles from $\left\{x_{k-1}^{(i)}, a_{k-1}^{(i)}\right\}_{i=1}^{L_k}$

3: **for** $i = 1 : L_k$ **do**

4:      Draw $x_k^{(i)}$ from $f_{k|k-1}(x_k | x_{k-1}^{(i)})$

5:      Compute $\{g_{k,j}(z_k | x_k^{(i)})\}_{j=1}^{N_f}$ according to Eq. (6.3)

6: **end for**

7: Compute $\{\alpha_{k,j}\}_{j=1}^{N_f}$

8: **for** $i = 1 : L_k$ **do**

9:      Compute $g_k(z_k | x_k^{(i)})$ according to Eq. (6.1)

10:     Assign the particle a weight $\omega_k^{(i)}$ according to Eq. (6.4)

11: **end for**

We will refer to the multi-feature particle filter with the proposed resampling procedure as MF-PFR and to the multi-feature particle filter with multinomial resampling (see Eq. 6.5) as MF-PF.

Finally, the best state at time k is derived using the Monte Carlo approximation of the expectation, as in Eq. (5.30). When the weights are updated online, we obtain the adaptive multi-feature particle filter (AMF-PFR) that is described by the Algorithm pseudo-code.

---

[9] The weights are said to be *unbalanced* when the weight of a single feature is very large and the weights of all the other features are small.

6.3.3 Feature reliability

The reliability of a feature quantifies its ability to represent a target based on its appearance as well as the ability to separate the target from the background and from other targets. We will discuss and compare various approaches for the computation of feature reliability.

6.3.3.1 *Distance to average*   The measure $\gamma^1_{k,j}$ (*distance to average*) for feature j is defined as

$$\gamma^1_{k,j} = \mathcal{R}\left(g_{k,j}(z_k|\hat{x}_k) - \left\langle g_{k,j}(z_k|x_k) \right\rangle \right), \tag{6.9}$$

where $\hat{x}_k$ is the state determined by the particle with maximum fused likelihood, defined as

$$\hat{x}_k = \arg\max_{x_k^{(i)}} \left\{ g_k(z_k|x_k^{(i)}) \right\}, \tag{6.10}$$

and $\left\langle g_{k,j}(z_k|x_k) \right\rangle$ is the average likelihood of feature j over the set of particles. $\mathcal{R}(.)$ is the ramp function

$$\mathcal{R}(x) = \begin{cases} x & \text{if } x > 0 \\ 0 & \text{otherwise} \end{cases}. \tag{6.11}$$

An alternative solution, here referred to as $\gamma^2_{k,j}$, substitutes $\hat{x}_k$ with the best particle selected separately by each feature j:

$$\hat{x}_{k,j} = \arg\max_{x_k^{(i)}} \left\{ g_{k,j}(z_k|x_k^{(i)}) \right\}. \tag{6.12}$$

6.3.3.2 *Centroid distance*   Feature reliability can also be estimated based on the level of agreement between each feature and the overall tracker result [12]. The contribution of each feature is a function of the Euclidean distance $\bar{E}_{k,j}$ between the centre of the best state estimated by feature j and the centre of the state obtained combining the features using the reliability scores at time $k-1$.

The corresponding reliability score $\gamma^3_{k,j}$ (*centroid distance*) is computed by smoothing $\bar{E}_{k,j}$ with a sigmoid function

$$\gamma^3_{k,j} = \frac{\tanh(-a\bar{E}_{k,j} + b) + 1}{2}, \tag{6.13}$$

where a, b are constants (fixed to $a = 0.4$ pixels^{-1} and $b = 3$ in the original paper [12]).

Note that this measure includes information extracted from the position estimates only, but does not account for other components of the target state, such as size and rotation.

6.3.3.3 Spatial uncertainty

To compute the likelihood as in Eq. (6.1), we estimate the mixture coefficients $\alpha_{k,j}$ based on each feature reliability. We weight the influence of each feature based on their spatial uncertainty [25].

The spatial uncertainty of a feature depends on the shape of the likelihood. In fact, to facilitate the task of the state estimator, the likelihood should be:

- smooth,

- unimodal (i.e. with a single peak)

- informative around the maximum (i.e. with a non-flat surface around the peak).

However, the likelihood may present multiple peaks and, in the worst case, their local maxima might be close to the predicted target position. This type of feature, when compared with a feature that presents a dominant peak in the likelihood, is more spatially uncertain.[10]

Figure 6.2 shows a sample scenario with background clutter where measuring the spatial uncertainty of a feature may help the tracker. While tracking the face, the colour-histogram information is ambiguous. In fact the colour of the box on the bottom left is similar to the skin colour and consequently the colour likelihood presents two major modes (Figure 6.2(b)). Orientation information is instead more discriminative (i.e. less spatially spread) on the real target (Figure 6.2(c)).

To estimate the spatial uncertainty, we analyse the eigenvalues of the covariance matrix $C_{k,j}$ of the particles $x_k^{(i)}$ weighted by the likelihood, and computed for each feature j at time k [26]. For illustrative purposes we now define $C_{k,j}$ for a 2D state, $x = (u, v)$. Then the 2×2 normalised covariance matrix is

$$
C_j =
\begin{bmatrix}
\dfrac{\sum_{i=1}^{L_k} l_j(u^{(i)}, v^{(i)})(u^{(i)} - \bar{u})^2}{\sum_{i=1}^{L_k} l_j(u^{(i)}, v^{(i)})} & \dfrac{\sum_{i=1}^{L_k} l_j(u^{(i)}, v^{(i)})(u^{(i)} - \bar{u})(v^{(i)} - \bar{v})}{\sum_{i=1}^{L_k} l_j(u^{(i)}, v^{(i)})} \\[4mm]
\dfrac{\sum_{i=1}^{L_k} l_j(u^{(i)}, v^{(i)})(u^{(i)} - \bar{u})(v^{(i)} - \bar{v})}{\sum_{i=1}^{L_k} l_j(u^{(i)}, v^{(i)})} & \dfrac{\sum_{i=1}^{L_k} l_j(u^{(i)}, v^{(i)})(v^{(i)} - \bar{v})^2}{\sum_{i=1}^{L_k} l_j(u^{(i)}, v^{(i)})}
\end{bmatrix}.
$$

(6.14)

For a more readable notation we have omitted k, and used $l_j(x_k)$ instead of $g_{k,j}(z_k|x_k)$. $\bar{x} = (\bar{u}, \bar{v})$ is the average over the samples weighted by the likelihood. The extension to the 5D state used in elliptic tracking is straightforward and a 5×5 covariance matrix is obtained.

---

[10] Note that this observation is similar to the considerations that were used to design the hybrid tracking algorithm described in Section 5.3.3.2.

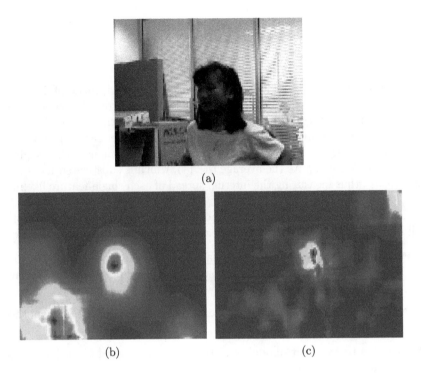

Figure 6.2 Comparison between the model-candidate likelihood of colour and orientation histograms in a head-tracking scenario. The target model is computed at track initialisation time. (a) Frame under analysis. (b) Spatial spread of the colour likelihood. (c) Spatial spread of the orientation likelihood. A reliability score measuring the spatial spread could improve the tracker performance by increasing the importance of the orientation histograms. IEEE © [26].

We can now define the uncertainty $U_{k,j}$ as

$$U_{k,j} = \sqrt[D]{\prod_{i=1}^{D} \lambda_{k,j}^{(i)}} = \sqrt[D]{|C_{k,j}|}, \qquad (6.15)$$

which is related to the volume of the hyper-ellipse having the eigenvalues

$$\{\lambda_{k,j}^{(i)}\}_{i=1}^{D}$$

as semi-axes. D is the dimensionality of the state space. The determinant $|.|$ is used instead of the sum of the eigenvalues to avoid problems related to state dimensions with different ranges (i.e. position versus size or orientation). The larger the hyper-volume, the larger the uncertainty of the corresponding feature about the state of the target.

The corresponding reliability score $\gamma_{k,j}^4$ (*spatial uncertainty*) is defined as

$$\gamma_{k,j}^4 = 1/U_{k,j}. \tag{6.16}$$

The importance of each feature is therefore the reciprocal of its uncertainty. $\gamma_{k,j}^4$ computes the average $(\bar{u}, \bar{v})$ for Eq. (6.15) from the particle states weighted by the fused likelihood using the reliability estimated at time $k-1$. This measures the likelihood spread compared with the tracker result.

A second version of the score $\gamma_{k,j}^5$ can be defined that uses the weights of each feature to compute the average $(\bar{u}_j, \bar{v}_j)$, which is substituted for $(\bar{u}, \bar{v})$ in Eq. (6.14), thus measuring the internal spread of each single-feature likelihood $g_{k,j}(.)$.

Both reliability estimations $\gamma_{k,j}^4$ and $\gamma_{k,j}^5$ based on particle filter sampling, allow us to assign different reliability scores to different targets in the scene:

- When the targets are far from each other in the state space, Eq. (6.15) determines the discriminative power of a feature in separating the target from the background.

- When the targets are close to each other, the two sets of hypotheses overlap.

Hence, due to the multi-modality of the likelihoods, the uncertainty defined by Eq. (6.15) increases.

6.3.4 Temporal smoothing

To impose temporal consistency and generate $\alpha_{k,j}$, the final score $\gamma_{k,j}$ undergoes filtering using the leaky integrator

$$\alpha_{k,j} = \tau\alpha_{k-1,j} + (1-\tau)\gamma_{k,j}, \tag{6.17}$$

where $\tau \in [0, 1]$ is the forgetting factor. The lower τ, the faster the update of $\alpha_{k,j}$. To satisfy Eq. (6.8), it is sufficient to enforce the condition

$$\sum_{j=1}^{N_f} \gamma_{k,j} = 1.$$

6.3.5 Example

As an example we analyse the time evolution of the five scores we discussed so far for a head-tracking task (Figure 6.3). The scores should reflect the following observations:

- When the head starts rotating, the contribution of the gradient should increase as the colour distribution changes significantly.

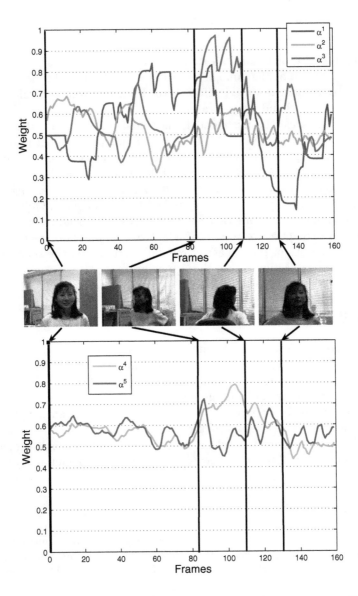

Figure 6.3 Comparison between different feature-weighting strategies. The plots show the evolution of the weights representing the relative importance of the shape information with respect to colour information during a complete revolution of the head (Key: α^1, α^2: distance to average; α^3: centroid distance; α^4, α^5: spatial uncertainty). IEEE © [26].

- When the head has again a frontal pose, the gradient contribution should decrease and approach its initial value.

From the plot of Figure 6.3 it can be seen that the score α^1 has a high variability, caused by the ramp function of Eq. (6.11). The likelihood evaluated in the best combined state is often lower than the average likelihood, thus resulting in $\gamma^1 = 0$ and a rapid variation of α^1. Unlike α^1, α^2 is not influenced by the ramp, since the likelihood is measured on the best particle of each feature separately. α^2 correctly increases the importance of the orientation histogram during the rotation. However, other variations are generated when no adaptation is expected. Similar considerations can be drawn for α^3; the high variability is not always motivated by real appearance changes. Before the head rotation, the two scores α^4 and α^5 behave similarly. Only α^4 has the expected adaptation profile that correctly follows the head rotation.

Extensive comparative results of adaptive multi-feature trackers and objective evaluations of the impact of reliability criteria on the tracking results are presented in Section A.3 in the Appendix.

6.4 SUMMARY

This chapter discussed the advantages of using multiple features in video tracking. An overview of different fusion strategies was presented based on differentiating approaches that combine the information at the tracker level and approaches that combine the information at the measurement level. We also discussed different reliability criteria and described an adaptive multi-feature tracker based on particle filtering.

While so far we have treated the problem of tracking a single target, in the next chapter we will describe how to deal with a variable number of targets.

REFERENCES

1. I. Leichter, M. Lindenbaum and E. Rivlin. A probabilistic framework for combining tracking algorithms. In *Proceedings of the IEEE Conference on Computer Vision and Pattern Recognition*, Vol. 2, Washington, DC, 2004, 445–451.

2. F. Moreno-Noguer, A. Sanfeliu and D. Samaras. Integration of conditionally dependent object features for robust figure/background segmentation. In *Proceedings of the International Conference on Computer Vision*, Washington, DC, 2005, 1713–1720.

3. G. Hua and Y. Wu. Measurement integration under inconsistency for robust tracking. In *Proceedings of the IEEE Conference on Computer Vision and Pattern Recognition*, New York, 2006, 650–657.

4. H. Veeraraghavan, P. Schrater and N. Papanikolopoulos. Robust target detection and tracking through integration of motion, color, and geometry. *Computer Vision and Image Understanding*, 103(2):121–138, 2006.

5. J. Sherrah and S. Gong. Fusion of perceptual cues using covariance estimation. In *Proceedings of the British Machine Vision Conference*, Nottingham, UK, 1999, 564–573.

6. A.D. Jepson, D.J. Fleet and T. El-Maraghi. Robust online appearance models for visual tracking. *IEEE Transactions on Pattern Analysis and Machine Intelligence*, 25(10):1296–1311, October 2003.

7. Y. Wu, G. Hua and T. Yu. Switching observation models for contour tracking in clutter. In *Proceedings of the IEEE Conference on Computer Vision and Pattern Recognition*, Madison, WI, 2003, 295–304.

8. K. Toyama and E. Horvitz. Bayesian modality fusion: Probabilistic integration of multiple vision algorithms for head tracking. In *Proceedings of the Fourth Asian Conference on Computer Vision (ACCV)*, Taipei, China, 2000.

9. J. Triesch and C. von der Malsburg. Democratic integration: Self-organized integration of adaptive cues. *Neural Computation*, 13(9):2049–2074, 2001.

10. M. Spengler and B. Schiele. Towards robust multi-cue integration for visual tracking. *Lecture Notes in Computer Science*, 2095:93–106, 2001.

11. P. Perez, J. Vermaak and A. Blake. Data fusion for visual tracking with particles. *Proceedings of the IEEE*, 92(3):495–513, 2004.

12. C. Shen, A. van den Hengel and A. Dick. Probabilistic multiple cue integration for particle filter based tracking. In *7th International Conference on Digital Image Computing (DICTA'03)*, Sydney, Australia, 2003, 309–408.

13. S. Khan and M. Shah. Object based segmentation of video using color, motion and spatial information. In *Proceedings of the IEEE Conference on Computer Vision and Pattern Recognition*, Kauai, HI, 2001, 746–751.

14. D. Kragic and H.I. Christensen. Cue integration for visual servoing. *IEEE Transactions on Robotics and Automation*, 17(1):18–27, 2001.

15. H. Kruppa and B. Schiele. Hierarchical combination of object models using mutual information. In *Proceedings of the British Machine Vision Conference*, Manchester, UK, 2001, 103–112.

16. J. L. Mundy and C.-F. Chang. Fusion of intensity, texture, and color in video tracking based on mutual information. In *Proceedings of the 33rd Applied Imagery Pattern Recognition Workshop*, Vol. 1, Los Alamitos, CA, 2004, 10–15.

17. Y. Wu and T.S. Huang. Robust visual tracking by integrating multiple cues based on co-inference learning. *International Journal of Computer Vision*, 58(1):55–71, 2004.

18. X. Zhong, J. Xue and N. Zheng. Graphical model based cue integration strategy for head tracking. In *Proceedings of the British Machine Vision Conference*, Vol. 1, Edinburgh, UK, 2006, 207–216.

19. S. Birchfield. Elliptical head tracking using intensity gradients and color histograms. In *Proceedings of the IEEE Conference on Computer Vision and Pattern Recognition*, Santa Barbara, CA, 1998, 232–237.

20. T.L. Liu and H.T. Chen. Real-time tracking using trust-region methods. *IEEE Transactions on Pattern Analysis and Machine Intelligence*, 26(3):397–402, 2004.

21. M. S. Arulampalam, S. Maskell, N. Gordon and T. Clapp. A tutorial on particle filters for online non-linear/non-Gaussian Bayesian tracking. *IEEE Transactions on Signal Processing*, 50(2):174–188, 2002.

22. J. Kittler, M. Hatef, R. W. Duin and J. Matas. On combining classifiers. *IEEE Transactions on Pattern Analysis and Machine Intelligence*, 20(3):226–239, 1998.

23. R. A. Jacobs. What determines visual cue reliability? *Trends in Cognitive Sciences*, 6(8):345–350, 2002.

24. M. Isard and A. Blake. CONDENSATION – conditional density propagation for visual tracking. *International Journal of Computer Vision*, 29(1):5–28, August 1998.

25. K. Nickels and S. Hutchinson. Estimating uncertainty in SSD-based feature tracking. *Image and Vision Computing*, 20(1):47–58, 2002.

26. E. Maggio, F. Smeraldi and A. Cavallaro. Adaptive multifeature tracking in a particle filtering framework. *IEEE Transactions on Circuits and Systems for Video Technology*, 17(10):1348–1359, 2007.

7

MULTI-TARGET MANAGEMENT

7.1 INTRODUCTION

The trackers presented in the previous chapters generally rely on an initial estimate of the target position in the image plane. This is an operational condition that is acceptable when developing a tracking algorithm in specific applications where the tracker can rely on user initialisation. When instead the application requires real-time tracking of a time-varying number of targets, the tracking algorithm needs to integrate automated initialisation and termination capabilities.

Most multi-target algorithms use detection-based tracking [1, 2] where, as discussed in Chapter 1, the multi-target tracking problem is solved in two steps:

- the *detection* of the objects of interests and

- the *association* of consistent identities with different instances of the same object over time.

In the second step, multi-target trackers compute the optimal association between sets of detections produced in subsequent frames and then use the association to produce an estimate of the number of targets $M(k)$ and their positions X_k. We can identify two main groups of multi-target algorithms, namely based on finding the nearest neighbour solution and based on solving linear assignement problems:

1. Algorithms based on *nearest neighbour* associate each trajectory with the closest measurement to the predicted position of the target [3].

2. Algorithms based on finding a global optimum by formulating data association as a *linear assignment* problem [4] offer principled solutions for track initialisation and termination. When the problem is cast over a window of frames, these algorithms can cope with objects failing to generate detections (observations), as in the case of short occlusions [5]. Among the methods of this class the *multiple-hypothesis tracker* (MHT) [6] propagates over time the best trajectory hypotheses, including new and disappearing targets, missing and false detections. As with other data association strategies, MHT propagates each association hypothesis via a single-target localisation method, often a Kalman Filter [6].

An important limitation of association algorithms of the second class is that the number of association hypotheses grows rapidly with the number of targets in the scene and the temporal observation window. To mitigate this problem, a *gating* procedure is often applied to discard less probable associations. More recently, to solve this problem, a novel framework based on Random Finite Sets (RFS) was proposed [7]. This framework offers the opportunity to perform smart approximations that result in linear complexity, as we will discuss later in this chapter.

7.2 MEASUREMENT VALIDATION

Given the trajectory $\mathbf{x}_{k-1}$ of a target (i.e. the set of state estimates up to frame $k-1$) and the new set of measurements Z_k at time k, we want to find which measurements are compatible with the expected motion of the target.

Measurement validation, also known as *gating*, is a common technique used to reduce the number of association hypotheses and consequently to reduce the computational cost. The main idea is that some measurements are not compatible with the expected motion of the target and therefore it is possible to discard them based on some validation criteria.

A common approach to measurement validation is:

- to predict the current state $\hat{x}_k$ from the past data (i.e. $\mathbf{x}_{k-1}$) and

- then to assess the compatibility of each measurement with this prediction.

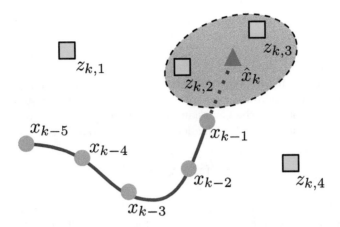

Figure 7.1 The measurements (green boxes) within the validation region (grey area) centred on the prediction (red triangle) are considered as possible associations with the trajectory (in blue).

This approach is particularly appealing for tracking algorithms where the prediction step is also part of the state-estimation procedure (e.g. algorithms based on Bayesian iterative filtering). Figure 7.1 shows a pictorial example of measurement validation.

From a Bayesian tracker perspective, the validation of a measurement $z \in Z_k$ should depend on the likelihood $p(z|\mathbf{z}_{k-1})$ of this measurement, conditional on the previous data associated with the trajectory. From the Markovian assumptions, it follows that

$$p(z|\mathbf{z}_{k-1}) = \int p(z|x_k)p(x_k|\mathbf{z}_{k-1})dx_k, \qquad (7.1)$$

where $\mathbf{z}_{k-1}$ are the measurements associated with the target in the previous time steps and that have been used to estimate the trajectory $\mathbf{x}_{k-1}$.

Typically, measurements with $p(z|\mathbf{z}_{k-1})$ that is higher than an appropriate threshold should be considered as possible associations for the trajectory data $\mathbf{x}_{k-1}$. When a Kalman filter is used then $p(z|\mathbf{z}_{k-1})$ is Gaussian and from Eq. (5.14) and Eq. (5.15) it follows that

$$p(z|\mathbf{z}_{k-1}) = \mathcal{N}(z|\hat{z}_k, S_k) = \mathcal{N}(z|G_k\hat{x}_k, G_k P_{k|k-1} G_k' + R_k), \qquad (7.2)$$

where $\hat{z}_k = G_k\hat{x}_k$ is the predicted measurement projected from the predicted state $\hat{x}_k$, $P_{k|k-1}$ is the covariance of the prediction probability

$$p(x_k|\mathbf{z}_{k-1}) = \mathcal{N}(x_k|\hat{x}_k, P_{k|k-1}),$$

and R_k is the measurement noise covariance.

Note that in some situations it might be necessary to validate measurements on non-consecutive time steps. In this case, the computation of $p(z|\mathbf{z}_{k-v})$, with $v > 1$, is similar to Eq. (7.2). Given the difference in time between the $(k - v)$th and kth steps and the linear state-propagation equation, we can calculate the predicted position $\hat{x}_k$ and the prediction covariance $P_{k|k-v}$. In practice, to save computational resources, we can avoid computing the expensive exponential of the Gaussian and validate those measurements with a Mahalanobis distance that is smaller than a threshold T, that is

$$d(z, \hat{z}_k) = \sqrt{(z_k - \hat{z}_k)'S_k^{-1}(z_k - \hat{z}_k)} < T. \qquad (7.3)$$

Since $d(z_k, \hat{z}_k)$ is χ^2-distributed, the validation threshold T is found by choosing a rejection percentile and by then looking up the value in the tables of the χ^2 cumulative distribution function.

In the case of multiple targets and multiple tracking hypotheses, the validation procedure should be repeated for each target and for each tracking hypothesis.

7.3 DATA ASSOCIATION

Given the results from an object detector, the tracking problem can be solved by linking over time the detections generated by the same target and then by using these detections to estimate its trajectory. Although the problem is NP-complex with the duration of the observation period, simplifications can be introduced to reduce the complexity.

7.3.1 Nearest neighbour

Let us consider the problem of associating one of the measurements in the set Z_k to a single target. The nearest neighbour is the simplest data association technique. Given the past state estimates $\mathbf{x}_{k-1}$ (i.e. the trajectory), the nearest neighbour method selects the measurement that is most compatible with the prior target evolution or, in other words, the closest measurement to the predicted measurement $\hat{x}_k$.

Note that the computational cost of finding the nearest neighbour can be reduced by pre-sorting the measurements. Instead of computing the distance between the prediction $\hat{x}_k$ and the measurements in Z_k one can organise the data in a *partitioning data structure* known as a kd-tree. A kd-tree is a binary tree where each non-leaf node is a k-dimensional point that splits the space into two subspaces. In our case, k is the dimensionality of the observation space.

Better performance is typically obtained with a balanced tree, a tree where the distance from the leaf nodes to the root node is similar for all the leaves. A

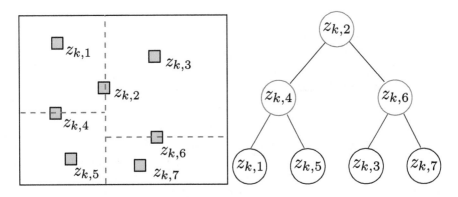

Figure 7.2 Sample of a kd-tree for nearest-neighbour measurement association. The balanced binary three (right) is formed by recursively splitting the observation space around the median measurment (left).

simple way to obtain a balanced tree, as the one in the right side of Figure 7.2, is to cycle through the dimensions of the measurement space and to use as a splitting point the median measurement for the current dimension, as in the left side of Figure 7.2. Once the kd-tree is formed, finding the nearest neighbour involves the following steps:

1. From the root, iteratively compare the predicted measurement $\hat{z}_k$ with the splitting plane and move to the branch the measurement belongs to, till reaching a leaf node.

2. Select the measurement associated to the leaf node as the current nearest neighbour.

3. Compute the distance d_{best} between $\hat{z}_k$ and the current nearest neighbour.

4. Move one step up the tree and compute the distance d_{s} between $\hat{z}_k$ and the splitting hyper-plane.

5. If $d_{\text{best}} \geq d_{\text{s}}$, look for the nearest neighbour on the other branch of the tree; otherwise repeat from step 4 until you reach the root.

Finding the nearest neighbour is $O(\log N_{\text{o}})$, where N_{o} is the number of measurements. This cost is cheaper than the naive extensive search strategy that is $O(N_{\text{o}})$. However, as the creation of the kd-tree requires extra computations, for small N_{o} extensive search may be faster than using a kd-tree.

Note that simple nearest-neighbour approaches require a set of heuristics to account for target birth and death, missing and false measurements.

The nearest-neighbour strategy can be easily extended to multiple targets by repeating the single assignment process for each target. This approach produces reasonable performance as long as the measurements generated by the targets are well separated in the measurement space. When this is not the case, a problem arises as there is the risk of assigning the same measurement to multiple targets.

7.3.2 Graph matching

To obtain a consistent identity for each target over time an optimised procedure based on graph-matching can be used [4, 5]. To this end, we can cast the problem of associating pre-existing trajectories $\mathbf{X}_{k-1}$ (or equivalently the existing assigned measurements $\mathbf{Z}_{k-1}$) with the current measurements Z_k as a linear assignment problem. We will distinguish three scenarios: two-frame matching with the same number of elements; two-frame matching with a different number of elements and multi-frame matching. These scenarios are described below.

7.3.2.1 Two-frame matching with the same number of elements   Let us assume that the two sets $\mathbf{Z}_{k-1}$ and Z_k have the same number of elements. Let us also define a cost function $c(\cdot)$ that evaluates the compatibility of a predicted target state with a measurement.

The objective is to find the assignment $\zeta_k : \mathbf{Z}_{k-1} \to Z_k$, a function that associates an element in $\mathbf{Z}_{k-1}$ with an element in Z_k that minimises

$$\min_{\zeta_k} \sum_{\mathbf{z} \in \mathbf{Z}_{k-1}} c(\mathbf{z}, \zeta_k(\mathbf{z})). \tag{7.4}$$

The optimal solution of this minimisation problem can be found using the Hungarian algorithm from Hopcroft and Karp [8]. The complexity of this algorithm for the tracking problem is cubic in the number of trajectories. Further savings can be achieved by applying gating before association (see Section 7.2). This is possible as the linear assignment problem can be interpreted in terms of a weighted bipartite graph such as the one shown in Figure 7.3(a).

A bipartite graph, also called a *bigraph*, is a graph where the set of vertices is composed by two disjoint subsets such that every edge in the graph connects a vertex in one set to a vertex in the other set. In our case the two sets are clearly identified in the existing trajectories $\mathbf{X}_{k-1}$ (or equivalently $\mathbf{Z}_{k-1}$) and the new measurements Z_k and the edges are weighted by the cost function. The final tracks are identified by the best set of edges generated by the path cover of the graph with the minimum cost (i.e. the combination of edges that gives the minimum cost sum). The number of edges in the graph influences the computational cost and these can be reduced by adding to the graph only those edges between nodes that pass the validation step (Figure 7.3(b)).

A common choice for the cost function $c(.)$ is the opposite of the measurement log probability

$$c(\mathbf{z}_{k-1}, z_k) = -\log p(z_k | \mathbf{z}_{k-1}), \tag{7.5}$$

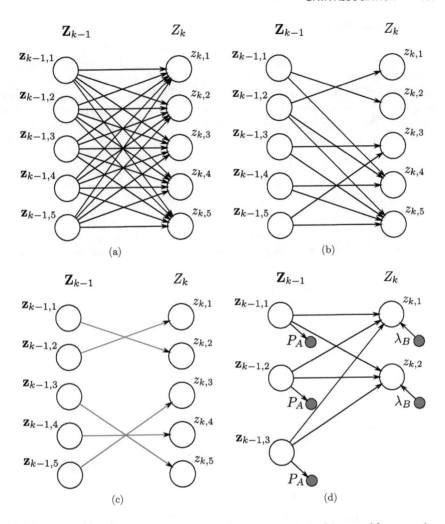

Figure 7.3 Graph matching. The graphs represent matching problems equivalent to the linear assignment between existing trajectories and measurements when the two sets have the same number of elements. (a) Full association graph. (b) Association graph after measurement validation. (c) Possible solution of the linear assignment problem. (d) Extended graph that accounts for unmatchable trajectories and measurements.

with $z_k \in Z_k$ and $\mathbf{z}_{k-1} \in \mathbf{Z}_{k-1}$. Therefore, assuming that the targets generate independent observations, solving the linear assignment problem is equivalent to maximising the logarithm of the association likelihood

$$p(Z_k | \zeta_k, \mathbf{Z}_{k-1}) = \prod_{\mathbf{z} \in \mathbf{Z}_{k-1}} p(\zeta_k(\mathbf{z}) | \mathbf{z}). \qquad (7.6)$$

7.3.2.2 Two-frame matching with a different number of elements   The
solution described so far is limited to the case where the number of trajectories
$\mathbf{X}_{k-1}$ is the same as the number of detections in Z_k. To relax the model to
cope with *unmatchable trajectories* and *unmatchable observations*, we can add
additional edges to the graph (Figure 7.3(d)). These edges (one per trajectory
and one per measurement) will work as slack variables in the optimisation
problem. The weights of the edges may have a double tracking meaning that
depends on the problem we are trying to solve for:

- If we assume a pre-defined and constant number of targets, unmatched
  trajectories and detections represent missing and false detections, respectively. Therefore, under the assumption of uniform clutter with Poisson
  intensity, the weights of the edges should be set to $-\log P_A$ and $-\log \lambda_B$,
  where P_A is the probability that a target fails to generate a detection and
  λ_B is the clutter Poisson rate modelling the average number of detections
  that are due to clutter.

- If there is a variable number of targets, unmatched trajectories and detections can represent terminated trajectories and new targets appearing
  in the scene. Therefore, under the assumption of uniform birth events
  with Poisson intensity, now P_A and λ_B are the death probability and
  new-born rate, respectively.

In both cases, the graph maximises the posterior

$$p(\zeta_k|Z_k, \mathbf{Z}_{k-1}) = \frac{1}{C}p(Z_k|\zeta_k, \mathbf{Z}_{k-1})p(\zeta_k|\mathbf{Z}_{k-1})$$

$$= \frac{1}{C}P_A^{N_A} \lambda_B^{N_B} \prod_{\mathbf{z}\in\hat{\mathbf{Z}}_{k-1}} (1 - P_A)p(\zeta_k(\mathbf{z})|\mathbf{z}), \qquad (7.7)$$

where C is a normalising factor and $\hat{\mathbf{Z}}_{k-1} \subseteq \mathbf{Z}_{k-1}$ contains the trajectories
that in the association hypotheses ζ_k are matched with a measurement. Note
that the weights of the edges connecting detections and trajectories are mul-
tiplied (summed in the logarithmic scale) by the factor $(1 - P_A)$ that defines
either the detection probability or the survival probability, respectively.

7.3.2.3 Multi-frame matching   Finally, the limitation of associating mea-
surements and trajectories frame-by-frame is that it is not possible to cope
with missing and false detections, and target birth and death at the same
time within the linear association framework.

Some of these shortcomings can be overcome by extending the graph to
encompass multiple frames [5]. The idea is to model possible missing detections by allowing for links between nodes of the graph from non-consecutive
frames. In the following, we explain in practical terms how to recursively form
the graph and how to find the optimal solution.

Let us assume that the best association between measurements in the frame range $[k-l, k-1]$ with $l > 1$ is known and that the solution may include links between non-consecutive frames (Figure 7.4(a)).

When the new measurements at time k become available, we add to the graph all those association hypotheses between the new measurements and all the other measurements in the previous $k - l$ frames that pass the validation procedure (Figure 7.4(b)). The resulting graph is not a bigraph as it contains nodes that have both input and output edges. To this end, we enforce a bipartitioning of the graph by splitting the nodes that have both input and output connections into two twin nodes, each one inheriting one of the two connection sets, as shown in Figure 7.3(c). Also, as in the graph of Figure 7.4(d), termination and initialisation edges can be included to model target birth and death. Then, algorithms like the Hungarian search ([8]) are applied to find the maximum path cover of this graph.

Unlike the two-frame graph, the solution of the assignment problem is optimal only when the cost function does not depend on past assignments (i.e. the cost depends on the z of the nodes connected by the edges only). This is not the case, for example, when the prediction uses target speed or acceleration estimates that are computed using previous association results. The problem is that the multi-frame associations are not definitive till the frame is pushed out of the time window $[k-l, k-1]$. In fact the solution at time k may over-ride some of the older associations that were used to compute speed and acceleration estimates. This bias can be corrected by the non-recursive algorithm proposed by Shafique et al. [5].

The probabilistic interpretation of the method is the maximisation of

$$p(\zeta_{k-l:k}|Z_{k-l:k}, \mathbf{Z}_{k-l-1}),$$

the joint association posterior conditioned over the measurements within the frame window and the old measurements. The cost of edges connecting possibly non-consecutive measurements becomes

$$c(\mathbf{z}_{k-v}, z_k) = -\log\left(P_M^{v-1} p(z_k|\mathbf{z}_{k-v})\right),$$

where P_M is the missing detection probability and $p(z_k|\mathbf{z}_{k-v})$ comes from forward-predicting the trajectory of v frames. Although the resulting algorithm can deal with target births, deaths and missing detections, it does not explicitly account for cluttered detections.

In the next section we describe a technique that, although more computationally intensive, generalises the association problem to its final level.

7.3.3 Multiple-hypothesis tracking

The multiple-hypothesis tracker (MHT) algorithm [6] extends the concepts presented in the previous sections and can be considered as the optimal solution to the data-association problem, under the assumption that the targets generate independent measurements. The MHT was originally designed for

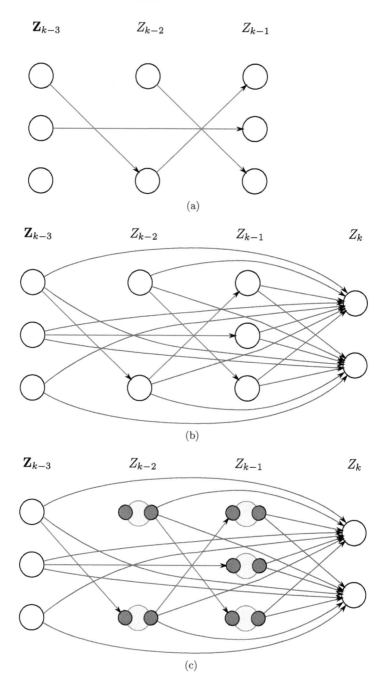

Figure 7.4 Multi-frame graph matching. The graphs represent a multi-frame data-association problem between tracking measurements. (a) Best association at time $k - 1$. (b) Possible associations at time k. (c) Enforcing the bipartitioning of the graph with twin nodes.

radar-based tracking applications and then used in vision-based applications [9]. Similarly to the multi-frame graph matching, the idea behind MHT is to delay a decision regarding the assignment of an old trajectory to a new measurement, until enough evidence becomes available.

7.3.3.1 *Global association*

In probabilistic terms, the goal of MHT is to find the global association hypothesis ζ_k (i.e. a set including all the associations up to frame k) that maximises the posterior $p(\zeta_k|\mathbf{Z}_k)$.

Let us assume that at step $k-1$ we have a set of plausible global association hypotheses and that ζ_{k-1} is one hypothesis in this set. Given the new set of measurements Z_k, we can spawn from each old global hypothesis a set of new global hypotheses by considering that each measurement in Z_k can:

1. be a spurious measurement (i.e. a false detection)

2. be associated to one of the trajectories postulated in ζ_{k-1} if it falls within its validation gate

3. be associated to a new trajectory.

Also, a trajectory postulated in ζ_{k-1} that is still alive at time $k-1$ can, at time k:

1. survive and generate a measurement

2. survive and remain undetected

3. disappear from the scene.

Given these conditions, MHT validates all the possible incremental hypotheses ζ_k that extend the parent global hypothesis ζ_{k-1} and computes the posterior probability of the novel global hypothesis $\zeta_k = \{\zeta_k, \zeta_{k-1}\}$ as

$$p(\zeta_k|\mathbf{Z}_k) = p(\zeta_k, \zeta_{k-1}|Z_k, \mathbf{Z}_{k-1}), \tag{7.8}$$

and using the Bayesian theorem

$$p(\zeta_k|\mathbf{Z}_k) = \frac{1}{C}p(Z_k|\zeta_k, \zeta_{k-1}, \mathbf{Z}_{k-1})p(\zeta_k|\zeta_{k-1}, \mathbf{Z}_{k-1})p(\zeta_{k-1}|\mathbf{Z}_{k-1}), \tag{7.9}$$

where the normalising factor C is obtained by summing the numerators over all the alive hypotheses at time k.

The first term $p(Z_k|\zeta_k, \zeta_{k-1}, \mathbf{Z}_{k-1})$ is the *likelihood of the current measurement*, given the hypothesis $\{\zeta_k, \zeta_{k-1}\}$ and the previous measurements $\mathbf{Z}_{k-1}$. Given the assignment ζ_k, a measurement $z_k \in Z_k$:

- can be generated from an existing target

- can be a spurious (false) measurement

- can be associated with a new-born target.

Therefore, assuming that the distribution of spurious measurements and of the measurements due to new-born targets and cluttered (false) measurements is uniform over a bounded observation volume V, then $p(Z_k|\zeta_k, \boldsymbol{\zeta}_{k-1}, \mathbf{Z}_{k-1})$ can be written as

$$p(Z_k|\zeta_k, \boldsymbol{\zeta}_{k-1}, \mathbf{Z}_{k-1}) = \frac{1}{V^{N_F+N_N}} \prod_{\mathbf{z} \in \boldsymbol{\zeta}_{k-1}(\mathbf{Z}_{k-1})} p(\zeta_k(\mathbf{z})|\mathbf{z}), \qquad (7.10)$$

where N_F and N_N are the number of false detections and new targets postulated by the hypothesis ζ_k, respectively; while the likelihood $p(\zeta_k(\mathbf{z})|\mathbf{z})$ of the measurement can be obtained from Eq. (7.2) with $\zeta_k(z) \in Z_k$.

The second term of Eq. (7.9), $p(\zeta_k|\boldsymbol{\zeta}_{k-1}, \mathbf{Z}_{k-1})$ is the *prior of the hypothesis* ζ_k, given the parent hypothesis $\boldsymbol{\zeta}_{k-1}$. To compute this probability we note that a target postulated by the hypothesis $\boldsymbol{\zeta}_{k-1}$ can either survive or disappear from the scene. Also, among the surviving targets, some may remain undetected. From these considerations and following the detailed derivation in [1], we obtain

$$p(\zeta_k|\boldsymbol{\zeta}_{k-1}, \mathbf{Z}_{k-1}) = \frac{N_F! N_N!}{M_k!} p(N_N) p(N_F) P_S^{N_S} (1-P_S)^{N-N_S} P_D^{N_D} (1-P_D)^{N-N_D-N_S},$$
$$(7.11)$$

where P_S is the probability of survival of a target.

Finally, the last term of Eq. (7.9) is the probability of the parent hypothesis $\boldsymbol{\zeta}_{k-1}$ computed at the previous step $k-1$.

From Eqs. (7.9), (7.10) and (7.11) and under the assumption that the number of false detections N_F as well as the number of new targets N_N are Poisson distributed, we obtain

$$p(\zeta_k|\mathbf{Z}_k) = \frac{1}{C'} \lambda_N^{N_N} \lambda_F^{N_F} \times$$
$$\times P_S^{N_S} (1-P_S)^{N-N_S} P_D^{N_D} (1-P_D)^{N-N_D-N_S} \times$$
$$\times p(\boldsymbol{\zeta}_{k-1}|\mathbf{Z}_{k-1}) \prod_{\mathbf{z} \in \boldsymbol{\zeta}_{k-1}(\mathbf{Z}_{k-1})} p(\zeta_k(\mathbf{z})|\mathbf{z}),$$

where λ_N and λ_F are the new arrival and false measurement rates defining the Poisson distributions.

The main problem with MHT is that *the number of tracking hypotheses grows exponentially over time*. Therefore, due to the limits on computational power and on memory, it is often not feasible to find the optimal solution of the association problem as formulated by the MHT. Hypothesis selection and pruning techniques can limit the growth in the number of hypotheses, while still finding a solution that is as close as possible to the optimum.

7.3.3.2 *Hypothesis propagation*

Two popular techniques for hypothesis propagation select a subset of hypotheses to be propagated.

The original method to reduce the number of hypotheses [6] is based on the observation that, in the case of well-separated targets, their tracks cannot be associated with the same measurements. For example, this happens when the measurement validation areas do not intersect each other. Given one old hypothesis ζ_{k-1} and the new measurements Z_k, we can form clusters of tracks and measurements that do not intersect in terms of associations. Given ζ_{k-1}, each cluster can be treated as a separate tracking problem. In this case, it is possible to form a hierarchy of independent, and therefore less computationally expensive, tracking problems that are solved separately.

A solution close to the optimum may be found by propagating over time only a handful of good hypotheses. Cox *et al.* [9] noted that the problem of finding the best ζ_k between the hypotheses spawned by one of the old ζ_{k-1} can be cast as a linear assigned problem similar to the two-frame problems we discussed in Section 7.3.2. Instead of the Hungarian search [8], which retrieves only the best hypotheses, Murty's algorithm [10] can be used to find the n best hypotheses. Also, it is possible to feed the algorithm with all the new association problems spawned from all the old hypotheses at the same time to find the globally best n hypotheses.

As final note, similarly to the other algorithms presented in the first part of this chapter, the MHT also assumes that each target generates one measurement (i.e. a detection) only and that the measurement is not shared among targets. This assumption is not valid for trackers based on traditional change detectors that generate blobs, as discussed in Section 1.2.1, when target proximity may induce multiple blobs to merge into a single measurement.

7.4 RANDOM FINITE SETS FOR TRACKING

As discussed in Section 5.2.2 and 5.3.2, the Bayesian recursion is a popular approach to filter noisy observations in single-target tracking [11, 12]. If the number of targets is fixed and known a priori, the extension of the Bayes recursion is trivial, but computationally intensive, as the dimensionality of the state space grows with the number of targets in the scene [1]. Moreover, none of the trackers described in the first part of this chapter is a natural extension of the single-target Bayes recursion to multi-target tracking. In fact, trackers like the MHT [6] apply independent Bayes filtering to each association hypothesis and not to the multi-target state X_k, thus reducing the filtering problem to a single-target one. In these filters, the estimate of the current number of targets $M(k)$ is a consequence of the selection of the best association hypothesis. A tracker based on a multi-target Bayes recursion should also filter $M(k)$ over time.

While the *appearance-based trackers* described in Section 1.3 model the uncertainty of the state and observation vectors x_k and z_k only, *detection-based multi-target trackers* must account for the time-varying number of targets in the scene and for the fact that an observation in Z_k may be due to clutter or that a target may fail to generate an observation. This requires extension of the uncertainty modelling to the cardinality of X_k and Z_k.

A Random Finite Set (RFS) provides a principled solution to the problem [13]. A RFS is a set where:

- the elements are random stochastic processes and

- the set cardinality is also a stochastic process.

Using the RFS framework it is possible to derive a multi-target Bayes recursion where uncertainty due to the time-varying number of targets, clutter and missing detections is accounted for.

To model the uncertainty on the multi-target state and measurement, a new formulation of the multi-target tracking problem and of the Bayes recursion which makes use of finite set statistics (FISS) and RFSs can be used [7]. This framework considers the multi-target state as a *single meta-target* and the observations as a single set of measurements of the *meta-sensor* [14]. The multi-target state can be therefore represented by a RFS, whose Bayesian propagation is similar to that of the single-target case (see Section 5.2.2).

Compared with existing solutions like Reid's MHT (see Section 7.3.3), trackers based on RFS and FISS present a more principled way to deal with the birth of new targets, clutter, missing detections and spatial noise. In fact, while most of the methods based on the enumeration of the data-association hypotheses apply spatial filtering to each single-target hypotheses, RFS-based methods can *integrate spatial and temporal filtering in a single framework.* Also, one of the approaches based on RFS, known as the probability hypothesis density (PHD) filter [14], presents a *complexity that is independent of the number of targets* in the scene, thus virtually solving the processing problems that usually affect multi-target tracking implementations.

Let Ξ_k be the RFS associated with the *multi-target state*

$$\Xi_k = S_k\left(X_{k-1}\right) \cup B_k\left(X_{k-1}\right) \cup \Gamma_k, \tag{7.12}$$

where:

- $S_k\left(X_{k-1}\right)$ denotes the RFS of *survived targets*

- $B_k\left(X_{k-1}\right)$ is the RFS of *targets spawned* from the previous set of targets X_{k-1}

- Γ_k is the RFS of the *new-born targets* [14].

The RFS Ω_k associated with the *measurement* is defined as

$$\Omega_k = \Theta_k\left(X_k\right) \cup \Upsilon_k, \tag{7.13}$$

where $\Theta_k(X_k)$ is the RFS modelling the measurements generated by the targets X_k, and Υ_k models *clutter* and *false alarms*.

Similarly to the single-target case:

- the dynamics of Ξ_k are described by the multi-target transition density $f_{k|k-1}(X_k|X_{k-1})$, while

- Ω_k is described by the multi-target likelihood $g_k(Z_k|X_k)$ [7].

The recursive equations equivalent to Eq. (5.12) and Eq. (5.13) are

$$p_{k|k-1}(X_k|Z_{1:k-1}) = \int f_{k|k-1}(X_k|X_{k-1})p_{k-1|k-1}(X_{k-1}|Z_{1:k-1})\mu(dX_{k-1})$$

$$(7.14)$$

and

$$p_{k|k}(X_k|Z_{1:k}) = \frac{g_k(Z_k|X_k)p_{k|k-1}(X_k|Z_{1:k-1})}{\int g_k(Z_k|X_k)p_{k|k-1}(X_k|Z_{1:k-1})\mu(dX_k)}, \qquad (7.15)$$

where μ is an appropriate dominating measure on $\mathcal{F}(E_s)$.[11]

As we saw in the previous section, the number of track hypotheses in MHT-like trackers grows exponentially over time. To limit computational cost and memory usage, a hypothesis pruning mechanism can be applied. Similarly, computing an approximation of the RFS Bayesian recursion can result in NP-complex algorithms. The next section describes a solution to this problem that is known as Probability Hypothesis Density filter.

7.5 PROBABILISTIC HYPOTHESIS DENSITY FILTER

Although it is possible to compute the recursion of Eq. (7.14) and Eq. (7.15) with Monte Carlo methods [14], as the dimensionality of the multi-target state X_k increases with the number of targets in the scene $M(k)$, the number of particles required grows *exponentially* with $M(k)$. For this reason, an approximation is necessary to make the problem computationally tractable.

To this end, instead of the posterior itself, the first-order moment of the multi-target posterior can be propagated [7]. The resulting filter is known as the PHD filter. As the dimensionality of the PHD is that of the single-target state, and does not change with the number of targets in the scene, the theoretical computational complexity is also independent (i.e. it is $O(1)$) of the number of targets. However, as we will show later in this chapter, efficient PHD estimation requires a number of computations that is usually *proportional* (and not exponential) to the number of targets [14].

---

[11] For a detailed description of RFSs, set integrals and the formulation of μ, please refer to [14] and [7].

The cost of the lower complexity is the *lack of information on the identity of the targets*. For particle-PHD a clustering step is necessary to associate the peaks of the PHD with target identities [15, 16]. Data association for the Gaussian mixture probability hypothesis density (GM-PHD) is easier, as the identity can be associated directly with each Gaussian [17,18]. However, these methods are limited by the linearity and Gaussianity assumptions on the transition and measurement models. Jump Markov models have been used to extend GM-PHD to manoeuvring targets [19, 20]. Also, implementations of the particle PHD have been tested on synthetic data [14,21], 3D sonar data [22], feature-point filtering [23] and groups-of-humans detection [24].

The PHD, $\mathcal{D}_\Xi(x)$, is the first-order moment of the RFS Ξ [7]. As the first-order statistics are computed on the set cardinality, the PHD is a function of the single-target state space E_s.

An alternative approach to define the PHD is via one of its properties. In fact, the PHD is the function that in any region $R \subseteq E_s$ of the state-space returns the expected number of targets in R, that is

$$E[|\Xi \cap R|] = \int_R \mathcal{D}_\Xi(x)\mathrm{d}x, \qquad (7.16)$$

where $|.|$ is used to denote the cardinality of a set.

In more practical terms, the PHD is a *multi-modal function whose peaks identify likely target positions*. By counting the number of peaks and evaluating the integrals around the local maxima, it is possible to derive a filtered estimate of the target locations.

If we denote $\mathcal{D}_{k|k}(x)$ as the PHD at time k associated with the multi-target posterior density $p_{k|k}(X_k|Z_{1:k})$, then the Bayesian iterative prediction and update of $\mathcal{D}_{k|k}(x)$ is known as the PHD filter [7].

The recursion of the PHD filter is based on three assumptions:

1. The targets evolve and generate measurements independently

2. The clutter RFS, Υ_k, is Poisson-distributed

3. The predicted multi-target RFS is Poisson-distributed.

While the first two assumptions are common to most Bayesian multi-target trackers ([1, 25–28]), the third assumption is specific to the derivation of the PHD update operator [27].

The *PHD prediction* is defined as

$$\mathcal{D}_{k|k-1}(x) = \int \phi_{k|k-1}(x,\zeta)\mathcal{D}_{k-1|k-1}(\zeta)d\zeta + \gamma_k(x), \qquad (7.17)$$

where $\gamma_k(.)$ is the intensity function of the new target birth RFS (i.e. the integral of $\gamma_k(.)$ over a region R gives the expected number of new objects per frame appearing in R). $\phi_{k|k-1}(x, \xi)$ is the analogue of the state transition probability in the single-target case

$$\phi_{k|k-1}(x, \xi) = e_{k|k-1}(\xi)f_{k|k-1}(x|\xi) + \beta_{k|k-1}(x|\xi), \qquad (7.18)$$

where $e_{k|k-1}(\xi)$ is the probability that the target still exists at time k and $\beta_{k|k-1}(.|\xi)$ is the intensity of the RFS that a target is spawned from the state ξ.

The *PHD update* is defined as

$$\mathcal{D}_{k|k}(x) = \left[p_M(x) + \sum_{z \in Z_k} \frac{\psi_{k,z}(x)}{\kappa_k(z) + \langle \psi_{k,z}, \mathcal{D}_{k|k-1} \rangle}\right] \mathcal{D}_{k|k-1}(x), \qquad (7.19)$$

where:

- $p_M(x)$ is the missing detection probability
- $\psi_{k,z}(x) = (1 - p_M(x))g_k(z|x)$, and $g_k(z|x)$ is the single-target likelihood defining the probability that z is generated by a target with state x
- $\langle f, g \rangle = \int f(x)g(x)dx$
- $\kappa_k(.)$ is the clutter intensity.

No generic closed-form solution exists for the integral of Eq. (7.17) and Eq. (7.19). Under the assumptions of Gaussianity and linearity one can obtain a filter that in principle is similar to the Kalman filter. This filter is known as the Gaussian Mixture PHD filter (GM-PHD) [17]. Given the limitations of the GM-PHD filter on the dynamic and observation models, in the next section we describe a more generic Monte Carlo implementation of the PHD recursion, known as the particle PHD filter [14].

7.6 THE PARTICLE PHD FILTER

A numerical solution for the integrals in Eq. (7.17) and Eq. (7.19) is obtained using a sequential Monte Carlo method that approximates the PHD with a (large) set of weighted random samples (Eq. 5.20).

Given the set

$$\left\{\omega_{k-1}^{(i)}, x_{k-1}^{(i)}\right\}_{i=1}^{L_{k-1}}$$

of L_{k-1} particles and associated weights approximating the PHD at time $k-1$ as

$$\mathcal{D}_{k-1|k-1}(x) \approx \sum_{i=1}^{L_{k-1}} \omega_{k-1}^{(i)} \delta\left(x - x_{k-1}^{(i)}\right), \qquad (7.20)$$

an approximation of the predicted PHD, $\mathcal{D}_{k|k-1}(x)$, with a set of weighted particles

$$\left\{ \tilde{\omega}_k^{(i)}, \tilde{x}_k^{(i)} \right\}_{i=1}^{L_{k-1}+J_k}$$

is obtained by substituting Eq. (7.20) into Eq. (7.17) and then applying separately importance sampling to both terms on the r.h.s.

In practice, first we draw L_{k-1} samples from the importance function $q_k(.|x_{k-1}^{(i)}, Z_k)$ to propagate the tracking hypotheses from the samples at time $k-1$; then we draw J_k samples from the new-born importance function $p_k(.|Z_k)$ to model the state hypotheses of new targets appearing in the scene. We will discuss the choice of $q_k(.|x_{k-1}^{(i)}, Z_k)$ and $p_k(.|Z_k)$ in Section 7.6.1.

The values of the weights $\tilde{\omega}_{k|k-1}^{(i)}$ are computed as

$$\tilde{\omega}_{k|k-1}^{(i)} = \begin{cases} \dfrac{\phi_k\left(\tilde{x}_k^{(i)}, x_{k-1}^{(i)}\right)\omega_{k-1}^{(i)}}{q_k\left(\tilde{x}_k^{(i)}|x_{k-1}^{(i)}, Z_k\right)} & i=1,...,L_{k-1} \\[4mm] \dfrac{\gamma_k(\tilde{x}_k^{(i)})}{J_k p_k\left(\tilde{x}_k^{(i)}|Z_k\right)} & i=L_{k-1}+1,...,L_{k-1}+J_k \end{cases} \tag{7.21}$$

Once the new set of observations is available, we substitute the approximation of $\mathcal{D}_{k|k-1}(x)$ into Eq. (7.19), and the weights

$$\left\{ \tilde{\omega}_{k|k-1}^{(i)} \right\}_{i=1}^{L_{k-1}+J_k}$$

are updated according to

$$\tilde{\omega}_k^{(i)} = \left[p_M(\tilde{x}_k^{(i)}) + \sum_{z \in Z_k} \frac{\psi_{k,z}(\tilde{x}_k^{(i)})}{\kappa_k(z) + C_k(z)} \right] \tilde{\omega}_{k|k-1}^{(i)}, \tag{7.22}$$

where

$$C_k(z) = \sum_{j=1}^{L_{k-1}+J_k} \psi_{k,z}(\tilde{x}_k^{(i)})\tilde{\omega}_{k|k-1}^{(j)}.$$

The particle PHD filter was originally designed to track targets generating *punctual observations* (radar tracking [7]). To deal with targets from videos, we have to adapt the dynamic and the observation models to account for the *area* of the target on the image plane. In the following we describe how to account for the two additional dimensions, the width and the height of a target, in the observation.

7.6.1 Dynamic and observation models

In order to compute the PHD filter recursion, the probabilistic model needs information regarding object dynamics and sensor noise. The information contained in the dynamic and observation models is used by the PHD filter to classify detections not fitting these priors as clutter.

Although many different types of model exist, to ease the explanation process, but without loss of generality, in this section we treat the case where the measurements are coming from an object detector [29–31] that approximate the target area in the image plane with a rectangle represented by centroid coordinates (u, v) and width and height $w \times h$. Therefore the observation generated by each target becomes

$$z_k = [u_{z_k}, v_{z_k}, w_{z_k}, h_{z_k}] \in E_o,$$

where the observation space is $E_o \subset \mathbb{R}^4$. Also, the single target state at time k is be defined as

$$x_k = [u_{x_k}, \dot{u}_{x_k}, v_{x_k}, \dot{v}_{x_k}, w_{x_k}, h_{x_k}] \in E_s,$$

where the four parameters, u_{x_k}, v_{x_k}, w_{x_k}, h_{x_k} are the filtered versions of the observation parameters, while $\dot{u}_{x_k}$ and $\dot{v}_{x_k}$, the speed components of the target, are fully estimated from the data. Note that the state space $E_s \subset \mathbb{R}^6$.

Under these assumptions, the magnitude of the motion of an object in the image plane depends on the distance of the object from the camera. Since acceleration and scale variations in the camera far-field are usually smaller than those in the near-field, we model the *state transition* $f_{k|k-1}(x_k|x_{k-1})$ as a first-order Gaussian dynamic with state dependent variances (SDV). This model assumes that each target has constant velocity between consecutive time steps, and acceleration and scale changes approximated by random processes with standard deviations proportional to the object size at time $k-1$, i.e.

$$x_k = \overbrace{\begin{bmatrix} A & 0_2 & 0_2 \\ 0_2 & A & 0_2 \\ 0_2 & 0_2 & I_2 \end{bmatrix}}^{G} x_{k-1} + \begin{bmatrix} B_1 & 0_2 \\ B_2 & 0_2 \\ 0_2 & B_3 \end{bmatrix} \begin{bmatrix} n_k^{(u)} \\ n_k^{(v)} \\ n_k^{(w)} \\ n_k^{(h)} \end{bmatrix}, \tag{7.23}$$

with

$$A = \begin{bmatrix} 1 & T \\ 0 & 1 \end{bmatrix}, \quad B_1 = w_{x_{k-1}} \begin{bmatrix} \frac{T^2}{2} & 0 \\ T & 0 \end{bmatrix},$$

$$B_2 = h_{x_{k-1}} \begin{bmatrix} 0 & \frac{T^2}{2} \\ 0 & T \end{bmatrix}, \quad \text{and} \quad B_3 = \begin{bmatrix} Tw_{x_{k-1}} & 0 \\ 0 & Th_{x_{k-1}} \end{bmatrix},$$

where 0_n and I_n are the $n \times n$ zero and identity matrices, and $\{n_k^{(u)}\}, \{n_k^{(v)}\}$, $\{n_k^{(w)}\}$ and $\{n_k^{(h)}\}$ are independent white Gaussian noises with standard deviations $\sigma_{n^{(u)}}, \sigma_{n^{(v)}}, \sigma_{n^{(w)}}$ and $\sigma_{n^{(h)}}$, respectively:

- $\{n_k^{(u)}\}$ and $\{n_k^{(v)}\}$ model the acceleration of the target, while

- $\{n_k^{(w)}\}$ and $\{n_k^{(h)}\}$ model the variation in size.

$T = 1$ is the interval between two consecutive steps ($k-1$ and k), which we take to be constant when the frame rate is constant. For simplicity, no spawning of targets is considered in the dynamic model.

The *observation model* is derived from the following considerations: when an object is partially detected (e.g. the body of a person is detected while his/her head is not detected), the magnitude of the error is dependent on the object size. Moreover, the error in the estimation of the target size is twice the error in the estimation of the centroid. This is equivalent to assuming that the amount of noise on the observations is proportional to the size of the targets, and that the standard deviation of the noise on the centroid is half that on the size.

To this end we define the single-target likelihood as a Gaussian SDV model, such that

$$g_k(z|x) = \mathcal{N}(z; Cx, \Sigma(x)), \tag{7.24}$$

where $\mathcal{N}(z; Cx, \Sigma(x))$ is a Gaussian function evaluated in z, centred in Cx and with covariance matrix $\Sigma(x)$. C is defined as

$$C = \begin{bmatrix} D & 0_{2 \times 3} \\ 0_{2 \times 4} & I_2 \end{bmatrix}, \quad \text{with } D = \begin{bmatrix} 1 & 0 & 0 \\ 0 & 0 & 1 \end{bmatrix}$$

and $0_{a \times b}$ is the $a \times b$ zero matrix. $\Sigma(x)$ is a diagonal covariance matrix, defined as

$$\text{diag}(\Sigma(x)) = \left[\frac{\sigma_{m^{(w)}}}{2} w_x, \frac{\sigma_{m^{(h)}}}{2} h_x, \sigma_{m^{(w)}} w_x, \sigma_{m^{(h)}} h_x \right].$$

Note that the SDV models described in Eq. (7.23) and Eq. (7.24) do not allow a closed-form solution of the PHD filter recursive equations (Eq. 7.17 and

Eq. 7.19). They require an algorithm such as the particle PHD that can handle generalised state-space models. In order to use GM-PHD [18] with SDV, an approximation based on the extended Kalman filter or on the unscented transformation is necessary [17]. The other functions that define the PHD recursion are defined below.

7.6.2 Birth and clutter models

In the absence of any prior knowledge about the scene, one can assume that the missing detection probability, $P_M(x)$, the probability of survival, $e_{k|k-1}(x)$, and the birth intensity $\gamma_k(x)$ are uniform over x. To this extent, we decompose $\gamma_k(x)$ as

$$\gamma_k(x) = \bar{s}b(x),$$

where $\bar{s}$ is the average number of birth events per frame and $b(x)$ is the probability density of a birth that we take to be uniform on the state space.

Similarly, we define the clutter intensity $\kappa_k(z)$ as

$$\kappa_k(z) = \bar{r}c(z),$$

and we assume the clutter density $c(z)$ to be uniform over the observation space. In Chapter 8 we will generalise birth and clutter models to non-uniform ones so that we can account for scene contextual information.

7.6.3 Importance sampling

In order to complete the definition of the particle PHD filter recursion we need to design the *importance sampling* functions for the Monte Carlo approximation. On the one hand, L_{k-1} old particles are propagated, as in CONDENSATION [11], according to the dynamics (i.e. $q_k(.|.) \propto f_{k|k-1}(.|.)$). On the other hand, drawing the J_k new-born particles is not straightforward, as the tracker should be able to reinitialise after an unexpected lost track or target occlusion.

When *prior knowledge* of the scene is available, the samples could be drawn from a localised $\gamma_k(.)$. However, no target birth would be possible in state regions with low $\gamma_k(.)$, as no particles would be sampled in these areas.

When *no prior knowledge* is available, drawing from a uniform non-informative $\gamma_k(.)$ (as in the one we use) would require too many particles to obtain a dense sampling in a 6D state space. To avoid this problem, we assume that the birth of a target happens in a limited volume around the measurements; thus we draw the J_k new-born particles from a mixture of Gaussians centred on the components of the set Z_k.

Hence, we define the importance sampling function for new-born targets $p_k(.|Z_k)$ as

$$p_k(x|Z_k) = \frac{1}{N(k)} \sum_{z \in Z_k} \mathcal{N}(x; [z, 0, 0], \Sigma_b(z)), \qquad (7.25)$$

where the elements of the 6×6 diagonal covariance matrix Σ_b are proportional to w_z and h_z, and are defined as

$$\text{diag}(\Sigma_b(z)) = [\sigma_{b,u} w_z, \sigma_{b,\dot{u}} w_z, \sigma_{b,v} h_z, \sigma_{b,\dot{v}} h_z, \sigma_{b,w} w_z, \sigma_{b,h} h_z].$$

Although drawing new-born particles from Eq. (7.25) allows dense sampling around regions where a birth is possible, the particle PHD recursion is also influenced by the resampling strategy used to select the most promising hypotheses. In the next section we discuss the resampling issues for the particle PHD filter that accounts for the different nature of the particles.

7.6.4   Resampling

At each iteration, J_k new particles are added to the old L_{k-1} particles. To limit the growth of the number of particles, a resampling step is performed after the update step. If classical multinomial resampling is applied ([14, 32]), then L_k particles are resampled with probabilities proportional to their weights from

$$\left\{ \tilde{\omega}_k^{(i)} / \hat{M}_{k|k}, \tilde{x}_k^{(i)} \right\}_{i=1}^{L_k + J_k},$$

where $\hat{M}_{k|k}$ is the total mass. This resampling procedure gives a greater chance of tracking hypotheses with high weight to propagate, thus pruning from the set unlikely hypotheses.

L_k is usually chosen to keep the number of particles per target constant. At each time step, a new L_k is computed so that $L_k = \rho \hat{M}_{k|k}$. Hence the computational cost of the algorithm grows *linearly* with the number of targets in the scene.

After resampling, the weights of

$$\left\{ \omega_k^{(i)}, x_k^{(i)} \right\}_{i=1}^{L_k}$$

are normalised to preserve the total mass.

Although multinomial resampling is appropriate for a single-target particle filter, this strategy poses a series of problems when applied to the PHD filter.

The prediction stage of the PHD (Eq. 7.17) generates two different sets of particles:

- The L_{k-1} particles propagated from the previous steps to model the state evolution of existing targets, with weights proportional to $\omega_{k-1}^{(i)}$.

- The remaining J_k particles modelling the birth of new targets, with weights proportional to the birth intensity $\gamma_k(.)$.

For multi-dimensional state spaces where birth events are very sparse (i.e. low $\gamma_k(.)$), the predicted weights $\tilde{\omega}_{k|k-1}^{(i)}$ of the new-born particles may be several orders of magnitude smaller than the weights of the propagated particles. In this case, as the probability of resampling is proportional to $\tilde{\omega}_k^{(i)}$ and thereby to $\tilde{\omega}_{k|k-1}^{(i)}$, it is possible that none of the new-born particles is resampled and propagated to the next step. Although the approximation of the PHD is still asymptotically correct, the birth of a new target also depends on combinatorial factors. Furthermore, when one or a few new-born particles are finally propagated, the PHD is not densely sampled around the new-born target, thus reducing the quality of the spatial filtering effect. Increasing the number of particles per target ρ, is not effective as the value of ρ should be very large and comparable with $1/\gamma_k(.)$.

To overcome this problem, a multi-stage pipeline can be constructed that *resamples the new-born particles independently from the others*. The idea is to separately apply multinomial resampling to the new-born particles by segregating them for a fixed number N_s of time steps. In this way one allows the weights to grow until they reach the same magnitude as those associated with particles modelling older targets. This multi-stage multinomial resampling strategy for the particle PHD filter is summarised by the Algorithm pseudo-code. Figure 7.5 shows an example of the multi-stage resampling pipeline when $N_s = 3$.

The *multi-stage multinomial resampling* preserves the total mass of whole set of particles $\hat{M}_{k|k}$ (this is a requirement of the PHD filter), as it preserves the total mass of the particles in each stage (see Step 7 and Step 11 of the Algorithm pseudo-code).

As we model proposal density

$$p_k\left(\tilde{x}_k^{(i)}|Z_k\right)$$

of the new-born particles with a mixture of Gaussians centred on the observations (Eq. 7.25), we can take

$$J_k = N(k) \cdot \tau,$$

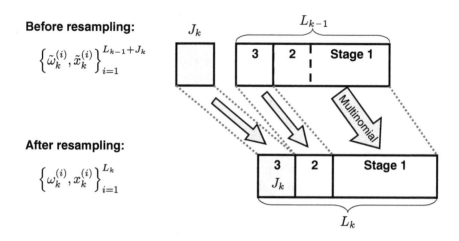

Figure 7.5   Schema of the multi-stage resampling strategy for the particle PHD filter (three-stage case). The J_k particles that model the birth of new targets are resampled separately from the older particles for a fixed number of time steps. IEEE © [33].

where τ is the number of new-born particles per observation. The overall computational cost of the algorithm now grows *linearly* with the number of targets X_k, and *linearly* with the number of observations Z_k.

7.6.4.1 *Example* In order to compare the multi-stage resampling strategy with the standard multinomial resampling, we analyse the statistics of the delay in the response of the filter produced by the resulting Monte Carlo approximations. To ensure that the difference is generated by the resampling only, we produce a synthetic scenario where the targets move according to the model described in Section 7.6.1. We fix one target in the centre of the scene and then we generate new targets uniformly distributed over the state space and according to a Poisson process. The two components of the speed of the new targets are uniformly drawn over the ranges $[-4\sigma_{b,\dot{u}}w_z, 4\sigma_{b,\dot{u}}w_z]$ and $[-4\sigma_{b,\dot{v}}h_z, 4\sigma_{b,\dot{v}}h_z]$, respectively. This also produces targets in regions of the state space with low density of new-born particles (Eq. 7.25). We collect the measurements Z_k for 1000 synthetic targets. We then give the measurements as input to the approximated PHD recursions using the two resampling strategies.

Table 7.1 shows the statistics related to the time delay in validating the new-born targets (expressed in frames), and the percentage of never-detected targets with respect to the speed ranges expressed as ratios between speed and object size. Higher ratios are associated with regions of the state space

---

Algorithm Multi-stage multinomial resampling

$$\left\{ \tilde{\omega}_k^{(i)}, \tilde{x}_k^{(i)} \right\}_{i=1}^{L_{k-1}+J_k} \rightarrow \left\{ \omega_k^{(i)}, x_k^{(i)} \right\}_{i=1}^{L_k}$$

1: **if** $k = 0$ **then**
2:      $S_i = 0 \quad \forall i = 1, \dots N_s$
3: **else if** $k \geq 1$ **then**
4:      $S_{N_s} = S_{N_s-1} + J_k$
5:      Compute the stage mass $\hat{M}_1 = \sum_{i=1}^{S_1} \tilde{\omega}_k^i$
6:      Compute the number of particles $\tilde{S}_1 = \hat{M}_1 \rho$
7:      Multinomially resample $\left\{ \tilde{\omega}_k^{(i)} / \hat{M}_1, \tilde{x}_k^{(i)} \right\}_{i=1}^{S_1}$ to get

$$\left\{ \omega_k^{(i)} = 1/\hat{M}_1, x_k^{(i)} \right\}_{i=1}^{S_1}$$

8:      **for** $j = 2 : N_s$ **do**
9:          Compute the stage mass $\hat{M}_j = \sum_{i=S_{j-1}+1}^{S_j} \tilde{\omega}_k^{(i)}$
10:         Compute the number of particles $\tilde{S}_j = \tilde{S}_{j-1} + \max\{\hat{M}_j \rho, S_j - S_{j-1}\}$
11:         Multinomially resample $\left\{ \frac{\tilde{\omega}_k^{(i)}}{\hat{M}_j}, \tilde{x}_k^{(i)} \right\}_{i=S_{j-1}+1}^{S_j}$ to get

$$\left\{ \omega_k^{(i)} = \frac{\hat{M}_j}{\tilde{S}_j}, x_k^{(i)} \right\}_{i=\tilde{S}_{j-1}+1}^{\tilde{S}_j}$$

12:      **end for**
13:      $L_k = \tilde{S}_{N_s}$
14:      $S_1 = \tilde{S}_1 + \tilde{S}_2$
15:      $S_i = \tilde{S}_{i+1} \quad \forall i = 2, \dots N_s - 1$
16: **end if**

---

where filtering is more difficult as the density of sampled particles (Eq. 7.25) is lower. Also, faster targets are more likely to leave the scene before the PHD filter manages to produce a target birth. The standard deviation of the filtering delay (Table 7.1) shows that the multi-stage resampling strategy has a beneficial effect in stabilising the behaviour of the filter (lower standard deviation). The higher average delay produced by multinomial resampling is due to those situations where none of the new-born particles is propagated to the next timestep. This is also confirmed by the higher percentage of missed targets produced by multinomial resampling.

A comparison between the multi-stage resampling strategy and the standard multinomial resampling is shown in Figure 7.6. The left column shows a delayed target birth (box) caused by the standard multinomial resampling. In this situation, dense sampling is made more difficult by the fast motion of the vehicle. Note that 30 frames of consecutive coherent detections are not enough to validate the target. Furthermore, when the first particles are resampled and propagated, the filtering result is poor due to the low number of samples available. Figure 7.6, right column, shows how the multi-stage

Table 7.1 Comparison of filtering response statistics between the standard multinomial resampling and the multi-stage multinomial resampling. The lower the initialisation delay and the missed targets percentage, the better.

	Initialisation delay	
	Multinomial	Multi-stage
Avg	11.2	5.1
Std dev	10.5	3.8

	Missed targets %	
Speed ratio	Multinomial	Multi-stage
0–0.05	23.8	9.9
0.05–0.1	29.4	8.7
0.1–0.15	45.6	16.2
0.15–0.2	50.0	27.5
0-0.25	**37.2**	**14.8**

resampling strategy improves the quality of the PHD approximation when new targets appear in the scene. The multi-stage multinomial resampling that uses the same birth intensity validates the track in four frames only, despite the motion of the target.

7.6.5 Particle clustering

After the resampling step, the PHD is represented by a set of weighted particles

$$\left\{ \omega_k^{(i)}, x_k^{(i)} \right\}_{i=1}^{L_k} \tag{7.26}$$

defined in the single-target state space.

An example of PHD approximated by particles is shown in Figure 7.7. The peaks of the PHD are on the detected vehicles and the mass $\hat{M}_{k|k} \approx 3$ estimates the number of targets. The local mass of the particles is larger where the tracking hypotheses are validated by consecutive detections. Note that although the set of particles carries information about the expected number of targets and their location in the scene, *the PHD does not hold information about the identity of the targets*. A clustering algorithm is required to detect

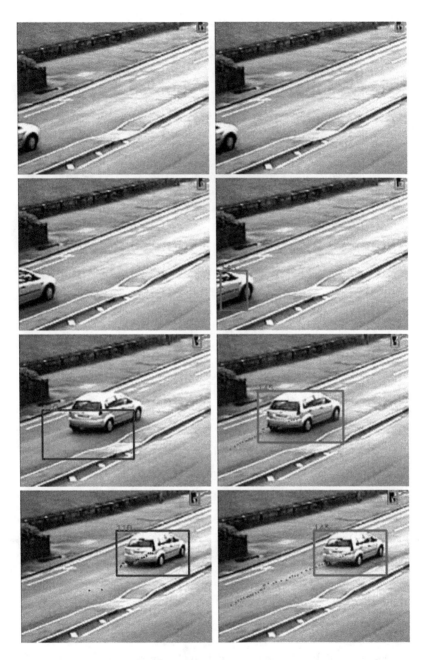

Figure 7.6 Sample tracking results using multinomial and multi-stage multi-nomial resampling (CLEAR-2007 dataset, sequence 102a03, frames 1354, 1359, 1385 and 1402). The multinomial resampling (left column) delays the initialisation of the track and introduces an error in the state estimation due to the low number of available samples. These behaviours are corrected by the multi-stage resampling strategy (right column). IEEE © [33].

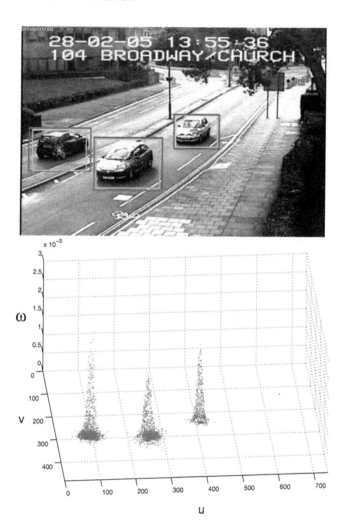

Figure 7.7 Visualisation of the particles approximating the PHD (before resampling) on the frame at the top when the vehicles are the targets (red boxes: input detections; green boxes: cluster centres). The PHD filter associates a cluster of particles (blue dots) with each target. The particle weights are plotted against the particle positions. The cumulative weight estimates the average number of targets in the scene. IEEE © [33].

the peaks of the PHD. These peaks define the set of candidate states

$$\bar{X}_k = \left\{ \bar{x}_{k,1}, \dots \bar{x}_{k,\bar{M}(k)} \right\}$$

of the targets in the scene and are the input of the data-association algorithm.

The information carried on by $\bar{X}_k$ is richer than that carried on by the original set of detections Z_k, as the elements of $\bar{X}_k$ are filtered in space and time by the PHD and include an estimate of the target velocity. Further information is also carried on by the total mass $\hat{M}_{k|k}$ estimating the expected number of targets in the scene. However, $\hat{M}_{k|k}$ may be composed of several clusters of particles with masses smaller than 1 and therefore the real number of clusters may be larger than $\hat{M}_{k|k}$.

To avoid underestimating the number of clusters, a top-down procedure [33] based on Gaussian mixture models (GMM), and inspired from the method applied by Vo *et al.* [17] to prune the components of the GM-PHD estimate, can be used. This procedure accounts for the new set of particles associated with target births, and updates the cluster parameters by means of an expectation maximisation (EM) algorithm. The intuitive reason for using Gaussian Mixture Model (GMM) is that both state dynamics and observation models are Gaussian, and therefore the clusters of particles also tend to be Gaussian-distributed.

The procedure works as follows: first, the set of clusters obtained at the previous step, $k - 1$, is augmented with new clusters initialised on the observations to model candidate new-born targets. Next, a hypothesis test is conducted to discard new clusters that are similar to old ones. The parameters of the remaining clusters are optimised by running EM on a set of $\check{L}_k = \rho_{GM} \hat{M}_{k|k}$ particles multinomially resampled [32] from

$$\left\{ \omega_k^{(i)} / \hat{M}_{k|k}, x_k^{(i)} \right\}_{i=1}^{\check{L}_k}.$$

Resampling is performed to obtain particles with uniform weights and also to reduce the computational cost of the EM recursion. After convergence, we discard small clusters (with mass below a threshold H) as they are usually associated with disappeared targets. Finally, we merge similar clusters according to a criterion based on hypothesis testing [17].

Let us define the parameters of the GMM at time k as

$$\theta_k = \left\{ \pi_{k,1}, \bar{x}_{k,1}, \Sigma_{k,1}, \ldots, \pi_{k,N_{c,k}}, \bar{x}_{k,N_{c,k}}, \Sigma_{k,N_{c,k}} \right\},$$

where $\pi_{k,i}$ is a weight coefficient of the mixture, $\bar{x}_{k,i}$ is the cluster centre, $\Sigma_{k,i}$ is the covariance matrix and $N_{c,k}$ is the number of clusters at time k. Given the cluster parameters ϑ_{k-1} at $k - 1$, the observation Z_k and the set of particles

$$\left\{ \omega_k^{(i)}, x_k^{(i)} \right\}_{i=1}^{L_k},$$

the clustering procedure that outputs the new set of clusters θ_k and the set of states $\bar{X}_k$ is detailed in the Algorithm box.

Algorithm Particle clustering

$$\left\{ \vartheta_{k-1}, Z_k, \left\{ \omega_k^{(i)}, x_k^{(i)} \right\}_{i=1}^{L_k} \right\} \rightarrow \left\{ \theta_k, \bar{X}_k \right\}$$

1: Multinomially resample $\left\{ \dfrac{\omega_k^{(i)}}{\hat{M}_{k|k}}, x_k^{(i)} \right\}_{i=1}^{L_k}$ to get $\left\{ \breve{\omega}_k^{(i)} = \dfrac{\hat{M}_{k|k}}{\breve{L}_k}, \breve{x}_k^{(i)} \right\}_{i=1}^{\breve{L}_k}$

2: $\forall z \in Z_k$ initialise a new cluster $\{1/N_k, [z, 0, 0], \Sigma_b(z)\}$ and add it to ϑ_N

3: $\forall$ clusters $c_j = \{\bar{x}_j, \Sigma_j\} \in \vartheta_N$ compute hypothesis test that $\bar{x}_{k-1,i} \in \vartheta_{k-1}$ is in the 99th percentile of c_j, and remove c_j from ϑ_N if $\exists i|$ the test is positive

4: Add the clusters to ϑ_{k-1} and, to obtain $\tilde{\vartheta}_k$, run EM till convergence on the particles obtained at step 1

5: Prune from $\tilde{\vartheta}_k$ the small clusters with $\tilde{\pi}_{k,j} < H$

6: Merge similar clusters (with the procedure defined in [17]) thus obtaining θ_k

7: Compute the cluster centres $\bar{X}_k$ according to $\bar{X}_k = \{\bar{x}_{k,i}, i = 1, \dots N_{c,k} | \pi_{k,i} \hat{M}_{k|k} < T_M\}$

7.6.6 Examples

The PHD can be used as a stand-alone algorithm to estimate the number of targets in the scene and their position (via clustering). However, its most common use is as a pre-processing step before data association. In this section we analyse partial and final tracking results with two different detectors, namely a change detector and a face detector.

Sample results of the PHD filter used to process the output of the *change detector* are shown in Figure 7.8. In this challenging situation generated by a sudden change in illumination, although the target size accuracy is not perfect, the heavy clutter is filtered by the smoothed values of the PHD (Figure 7.8(a)-(c)). Furthermore, Figure 7.9 shows the comparison between the multiple target tracker based on the PHD filter (PHD-MT) with a multiple target tracker (MT) that performs data-association directly on the measurement Z_k. In both cases the data-association algorithm is a sliding-window multi-frame graph-matching procedure, as the one described in Section 7.3.2. In cases when a target generates noisy observations, the spatial smoothing produced by the PHD filter facilitates data association preventing an identity switch on the same target (Figure 7.9, third row, the pedestrian in the centre of the scene).

To demonstrate the flexibility and modularity of the proposed multi-target framework, we show the results obtained when substituting the change detector with a *face detector* [31]. Figure 7.10 shows a comparison of the results obtained with and without the use of the PHD filter on the detected faces. When false detections are processed, the mass of the PHD starts growing around them. Multiple coherent and consecutive detections are necessary to

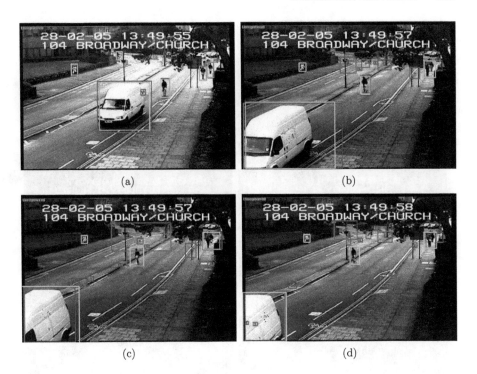

(a)   (b)

(c)   (d)

Figure 7.8 Comparison of tracking results on an outdoor surveillance scenario between raw object detections (colour-coded in red) from a background subtraction module and PHD filter output (colour-coded in green). Several false detections are filtered by the PHD (b, c, d). IEEE © [33].

increase the mass to a level greater than T_M. In fact the mass of the particles around a target initially depends on the birth $\gamma_k(.)$ (see Eqs. 7.17 and 7.21). Then, if the detections are temporally coherent with the transition and observation models defined by the two functions $\phi_{k|k-1}(x,\xi)$ and $\psi_{k,z}(x)$ the mass approaches a value of 1. The clutter model $\kappa_k(z)$ instead counteracts the increase of the mass (see Eqs. 7.19 and 7.22). For this reason, when the clutter is not persistent, the PHD filter removes it. As mentioned earlier, due to the trade-off between clutter removal and response time, the drawback of this filtering is a slower response in accepting the birth of a new target.

In addition to the above, Figure 7.11 shows how the combination of PHD filtering with the graph-matching-based data association is able to recover the identity of faces after a total occlusion: in the third and fourth row, although a face is occluded by another person, data association successfully links the corresponding tracks. Finally, the results of PHD-MT compared with MT shows that two false tracks on the shirt of one of the targets are removed by the PHD-MT only.

Further results and discussions are available in Section A.4 in the Appendix.

(a)   (b)

Figure 7.9 Comparison of tracking results between the multi-target tracker with (PHD-MT) and without (MT) PHD filter. (a) MT results. (b) PHD-MT results. False tracks due to clutter are removed by PHD-MT. IEEE © [33].

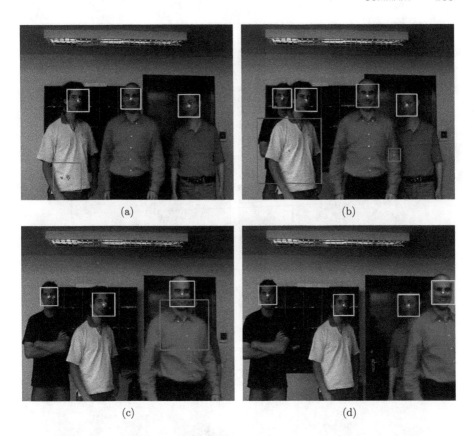

Figure 7.10 Comparison of tracking results in a face-tracking scenario between raw object detections (colour-coded in red) and PHD filter output (colour-coded in green). Several false detections are filtered by the PHD (a, b, c). IEEE © [33].

7.7 SUMMARY

In this chapter we discussed the problem of tracking a time-variable number of targets in a video. We first introduced three data-association methods that address different tracking challenges such as target initialisation and termination, clutter, spatial noise and missing detections. The second part of the chapter was dedicated to a novel tracking paradigm based on random finite set (RFS) and the probability hypothesis density (PHD) filter. We discussed how to adapt a filter based on RFS to real-world video-tracking scenarios.

Unlike the single-target particle filter, the multi-target PHD filter generates particles with two different purposes:

- to propagate the state of existing targets
- to model the birth of new ones.

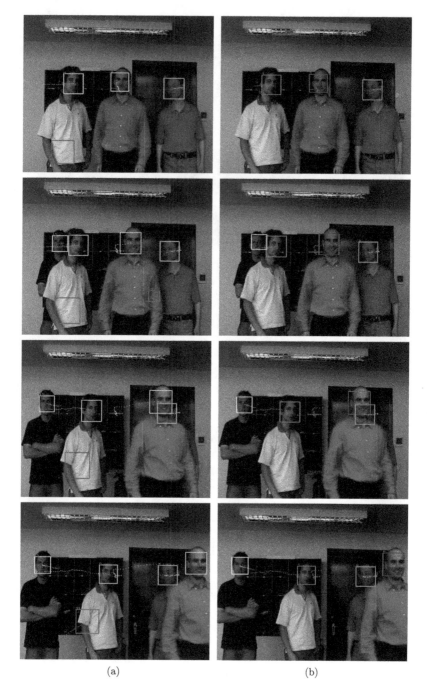

(a)   (b)

Figure 7.11   Comparison of tracking results between the multi-target tracker with (a) and without (b) Particle PHD filtering. The tracker with the PHD filter successfully recovers the faces after a total occlusion, without generating false tracks. IEEE © [33].

A multi-stage resampling strategy accounts for the different nature of the particles and, compared to the multinomial strategy improves the quality of the Monte Carlo estimation from a tracking perspective. As for the dynamic and observation models, state dependent variances (SDV) are used to account for the size of the targets. These models are not limited to the PHD recursion and can be implemented in any Bayesian recursive algorithms that can handle SDV.

In the next chapter we discuss how to introduce contextual information in the tracking framework by learning mixture models of clutter and birth densities.

REFERENCES

1. Y. Bar-Shalom and T.E. Fortmann. *Tracking and Data Association*. New York, Academic Press, 1988.

2. S.W. Joo and R. Chellappa. Multiple-hypothesis approach for multiobject visual tracking. *IEEE Transactions on Image Processing*, 16(11):2849–2854, 2007.

3. X. Rong Li and Y. Bar-Shalom. Tracking in clutter with nearest neighbor filters: analysis and performance. *IEEE Transactions on Aerospace and Electronic Systems*, 32(3):995 1010, 1996.

4. R.E. Burkard and E. Cela. Linear assignment problems and extensions. In *Handbook of Combinatorial Optimization*, Dordrecht, Kluwer Academic Publishers, 1999, 75–149.

5. K. Shafique and M. Shah. A noniterative greedy algorithm for multiframe point correspondence. *IEEE Transactions on Pattern Analysis and Machine Intelligence*, 27(1):51–65, 2005.

6. D. Reid. An algorithm for tracking multiple targets. *IEEE Transactions on Automatic Control*, 24(6):843–854, December 1979.

7. R. Mahler. A theoretical foundation for the Stein-Winter Probability Hypothesis Density (PHD) multitarget tracking approach. In *Proceedings of the 2002 MSS National Symposium on Sensor and Data Fusion*, Vol. 1, San Antonio, TX, 2000.

8. J.E. Hopcroft and R.M. Karp. An $n^{2.5}$ algorithm for maximum matchings in bipartite graphs. *Siamese Journal of Computing*, 2(4):225–230, December 1973.

9. I.J. Cox and S.L. Hingorani. An efficient implementation of Reid's multiple hypothesis tracking algorithm and its evaluation for the purpose of visual tracking. *IEEE Transactions on Pattern Analysis and Machine Intelligence*, 18(2):138–150, 1996.

10. K.G. Murty. An algorithm for ranking all the assignments in order of increasing cost. *Operations Research*, 16(3):682–687, 1968.

11. M. Isard and A. Blake. CONDENSATION – conditional density propagation for visual tracking. *International Journal of Computer Vision*, 29(1):5–28, August 1998.

12. M. Isard and A. Blake. CONDENSATION: Unifying low-level and high-level tracking in a stochastic framework. *Lecture Notes in Computer Science*, 1406: 893–908, 1998.

13. R. Mahler. Multi-target Bayes filtering via first-order multi-target moments. *IEEE Transactions on Aerospace and Electronic Systems*, 39(4), 1152–1178, 2003.

14. B.N. Vo, S.R. Singh and A. Doucet. Sequential Monte Carlo implementation of the PHD filter for multi-target tracking. In *Proceedings of the International Conference on Information Fusion*, Vol. 2, Cairns, Australia, 2003, 792–799.

15. D.E. Clark and J. Bell. Data association for the PHD filter. In *Proceedings of the Second International Conference on Intelligent Sensors, Sensor Networks and Information Processing*, Melbourne, Australia, 2005, 217–222.

16. K. Panta, Ba-Ngu Vo and S. Singh. Novel data association schemes for the probability hypothesis density filter. *IEEE Transactions on Aerospace and Electronic Systems*, 43(2):556–570, 2007.

17. B.-N. Vo and W.-K. Ma. The Gaussian mixture probability hypothesis density filter. *IEEE Transactions on Signal Processing*, 54(11):4091–4104, 2006.

18. D.E. Clark, K. Panta and VoBa-Ngu. The GM-PHD filter multiple target tracker. In *Proceedings of the International Conference on Information Fusion*, Quebec, Canada, 2006, 1–8.

19. B.-N. Vo, A. Pasha and H.D. Tuan. A Gaussian mixture PHD filter for nonlinear jump markov models. In *Proceedings of the IEEE Conference on Decision and Control*, San Diego, CA, 2006, 3162–3167.

20. A. Pasha, B. Vo, H.D. Tuan and W.-K. Ma. Closed form PHD filtering for linear jump markov models. In *Proceedings of the International Conference on Information Fusion*, Florence, Italy, 2006, 1–8.

21. H. Sidenbladh and S. Wirkander. Tracking random sets of vehicles in terrain. In *Proceedings of the IEEE Workshop on Multi-Object Tracking*, Madison, WI, 2003, 98.

22. D.E. Clark, I.T. Ruiz, Y. Petillot and J. Bell. Particle PHD filter multiple target tracking in sonar images. *IEEE Transactions on Aerospace and Electronic Systems*, 43(1):409–416, 2006.

23. N. Ikoma, T. Uchino and H. Maeda. Tracking of feature points in image sequence by SMC implementation of PHD filter. In *Proceedings of the SICE Annual Conference*, Vol. 2, Sapporo, Japan, 2004, 169–170.

24. Y.D. Wang, J.K. Wu, A.A. Kassim and W.M. Huang. Tracking a variable number of human groups in video using probability hypothesis density. In *Proceedings of the IEEE International Conference on Pattern Recognition*, Vol. 3, Hong Kong, China, 2006, 1127–1130.

25. J. Vermaak, S. Godsill and P. Perez. Monte Carlo filtering for multi-target tracking and data association. *IEEE Transactions on Aerospace and Electronic Systems*, 41(1):309–332, 2005.

26. C. Hue, J.-P. Le Cadre and P. Prez. Tracking multiple objects with particle filtering. *IEEE Transactions on Aerospace and Electronic Systems*, 38(3):791–812, July 2002.

27. A. Doucet, B. Vo, C. Andrieu and M. Davy. Particle filtering for multi-target tracking and sensor management. In *Proceedings of the International Conference on Information Fusion*, Vol. 1, Annapolis, MD, 2002, 474–481.

28. S. Sarkka, A. Vehtari and J. Lampinen. Rao-Blackwellized particle filter for multiple target tracking. *Information Fusion*, 8(1):2–15, January 2007.

29. C. Stauffer and W.E.L. Grimson. Learning patterns of activity using real-time tracking. *IEEE Transactions on Pattern Analysis and Machine Intelligence*, 22(747–757), 2000.

30. A. Cavallaro and T. Ebrahimi. Interaction between high-level and low-level image analysis for semantic video object extraction. *EURASIP Journal of Applied Signal Processing*, 6:786–797, June 2004.

31. P. Viola and M.J. Jones. Robust real-time face detection. *International Journal of Computer Vision*, 57(2):137–154, May 2004.

32. M.S. Arulampalam, S. Maskell, N. Gordon and T. Clapp. A tutorial on particle filters for online non-linear/non-Gaussian Bayesian tracking. *IEEE Transactions on Signal Processing*, 50(2):174–188, 2002.

33. E. Maggio, M. Taj and A. Cavallaro. Efficient multi-target visual tracking using random finite sets. *IEEE Transactions on Circuits Systems and Video Technology*, 18(8), 1016–1027, 2008.

8

CONTEXT MODELING

8.1 INTRODUCTION

In the previous chapters we discussed algorithms that operate independently from the scene context in which the objects move. As the behaviour of the objects depends on the structure of the scene, there is additional contextual information that can be expolited to improve the performance of a video tracker.

Contextual information can be acquired as an *external input* or can be *learnt online* by the tracker itself based on observations. One can receive information from an external source, such as *maps* (geographical information), or from *scene annotation* (i.e. the position of doors, traffic lights and trees). The data related to scene annotation can be either gathered automatically by means of image segmentation and object classifiers, or it can be provided interactively by the user. Moreover, the result of the tracker itself can be used to estimate contextual information by inferring the position of contextual object from the *analysis of the trajectories*. This estimation can be either fully automatic or can be generated interactively through (limited) user interaction.

Video Tracking: Theory and Practice. Emilio Maggio and Andrea Cavallaro
© 2011 John Wiley & Sons, Ltd

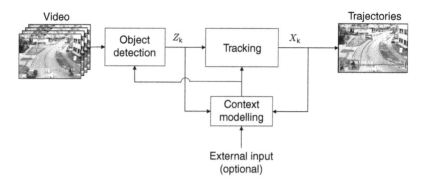

Figure 8.1 Context as information source for video tracking.

In this chapter we address the question of how to define or to learn the scene layout and in turn to help the tracker (Figure 8.1). In particular, we discuss how to use information about the scene to reduce the uncertainty in the state estimation.

8.2 TRACKING WITH CONTEXT MODELLING

8.2.1 Contextual information

Figure 8.2 shows two examples illustrating areas of the field of view of the cameras that are characterised either:

- by entry and exit areas (where the track will either start or finish) that are associated to the initialisation and the termination of a track, respectively, or

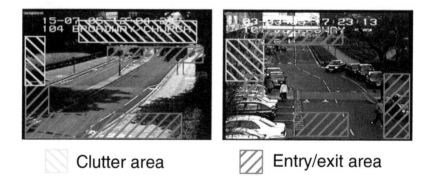

Figure 8.2 Example of contextual information to be used by the tracker. Image regions where clutter events (blue) and target birth events (violet) are likely to happen. Unmarked images from i-Lids (CLEAR 2007) dataset; reproduced with permission of HOSDB.

- by clutter areas (where vegetation is likely to generate false targets or to hide real targets).

These areas, if known, can be used to modify the behaviour of the tracker itself.

An interesting property of the RFS framework described in the previous chapter, and consequently of the PHD filter, is that it can use scene contextual information by allowing state-space-dependent models of *birth* and *clutter* [1]. The PHD filter can *account for scene context by varying its filtering strength according to the hypothetical state of a target.* This is possible as the birth intensity $\gamma_k(x)$ may depend on the state x and the clutter intensity $\kappa_k(z)$ may depend on the observation z. The acquisition of event and clutter information for intensity learning can be based on the analysis of the output of the tracker and of the detector, as discussed in the following sections.

The goal is to extract meaningful information that is representative of the densities of interest (e.g. birth and clutter) in order to estimate them and to introduce dependencies on the state and observation spaces. This is particularly relevant to real-world tracking as the intensity of birth and clutter events usually depends on the field of view of the camera. With reference to Figure 8.2, in image areas where new targets are likely to appear, the target birth model should account for spatial variability in the scene and allow the filter to reduce temporal smoothing over these locations. Likewise, the strength of the filtering should be increased in image areas where a detector based on background subtraction is expected to fail (e.g. on waving vegetation). For these reasons, the density of a birth and clutter event should depend on the scene contextual information.

8.2.2   Influence of the context

As considering spatial dependencies requires the introduction of additional parameters in the filter, the complexity is usually limited either by working with synthetic data, whose model matches exactly that used in the filter [2], or by using uniform distributions [3–6]. The latter case leads to a filter that is independent from the scene context. This is a common solution in many multi-target tracking algorithms and data-association methods (see Chapter 7) that assume that targets may appear everywhere in the scene with uniform probability and that the detectors may produce errors uniformly in any positions in the image.

The PHD filter and the full multi-target Bayes filter can *validate in space and time* the output of an object detector (see Section 7.5). The propagation model defined in Eq. (7.17) and Eq. (7.19) offers several degrees of freedom that help to tune the filtering properties depending on the tracking scenario at hand. As mentioned above, a simple solution to model clutter and birth intensities is to consider $\gamma_k(x)$ and $\kappa_k(z)$ uniform in x and z, respectively. In this case, the absolute values of the intensities representing the average number of birth and clutter events per frame are the only parameters to

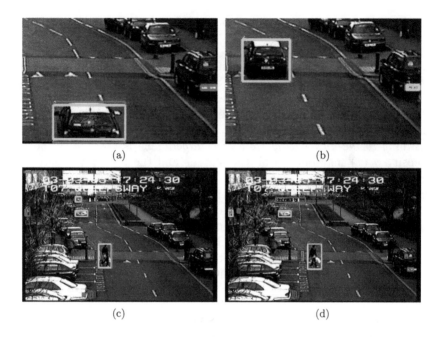

(a)   (b)

(c)   (d)

Figure 8.3   Examples of failures of a PHD-based tracker that uses uniform birth and clutter intensities on the Queensway scenario (QW) of the CLEAR-2007 dataset. The use of uniform densities leads to insufficient filtering that results in a false track on the number plate of the car (a) and (b) and on excessive filtering that results in a missed target (black car) (c) and (d) (red: detections from a background subtraction algorithm; green: the PHD output). IEEE © [1].

choose. However, targets tend to appear in specific locations of the image. Likewise, detection errors are more likely to occur in some areas of the image. For this reason, solutions that require a *compromise between the various image regions* may produce suboptimal results. For example, the filtering can be either too strong or too weak:

- If it is too stong, then correct detections are filtered out. Figure 8.3(c) and (d) shows an example of results where filtering is too strong. A car in the far field is detected for a few frames. Although the probability of appearance of a vehicle is high in that image region, the PHD filters out the detections, thus losing the track.

- If it is too weak, then persistent clutter is not removed. Figure 8.3(a) and (b) shows an example of PHD-based tracker results where the filtering is too weak. Spatially consistent detections caused by an illumination change are produced by the number plate of the car. Although this object is in a position where a new target is unlikely to be born, the filtering effect of the PHD is not strong enough and the tracker generates a false track.

We therefore need to learn an appropriate space-variant intensity model for birth and clutter.

8.3 BIRTH AND CLUTTER INTENSITY ESTIMATION

Learning intensity functions reduces to a density estimation problem by decomposing the birth and clutter intensities, $\gamma_k(x)$ and $\kappa_k(z)$, as

$$\gamma_k(x) = \bar{s}_k p_k(x|b), \tag{8.1}$$

and

$$\kappa_k(z) = \bar{r}_k p_k(z|c), \tag{8.2}$$

where $\bar{s}_k$ and $\bar{r}_k$ are the average birth/clutter events *per frame* and $p_k(x|b)$ and $p_k(z|c)$ are the birth/clutter event distributions defined over the state and observation spaces, respectively. For simplicity, we assume that the two intensities are stationary and therefore we drop the temporal subscript k. We can include this information in the PHD filtering model by computing approximated versions of $\bar{s}$, $\bar{r}$, $p(x|b)$ and $p(z|c)$.

The computation of $\bar{s}$ and $\bar{r}$ simply requires the cardinality of the event training sets and it is computed as the *average* over all the frames; $p(x|b)$ and $p(z|c)$ instead require density estimation in 6D and 4D spaces, respectively.

In the next part of this section we will describe how to tune the filter by learning non-uniform models of birth and clutter densities.

8.3.1 Birth density

For the estimation of the density $p(x|b)$ of Eq. (8.1), evidence of *birth events*

$$X_b = \{x_{b,i}\}_{i=1}^{M_b},$$

representing the initial locations, sizes and velocities of the objects appearing in the scene is collected based on an analysis of the tracker output (see Figure 8.4) according to the following set of rules:

- As long trajectories with a coherent motion are unlikely to be generated by clutter, we use their initial state to model *birth* events. We add to the set of birth states X_b all the starting positions from trajectories whose lifespan is longer than a certain minimum period (e.g. 3 seconds for a typical outdoor surveillance scenario).

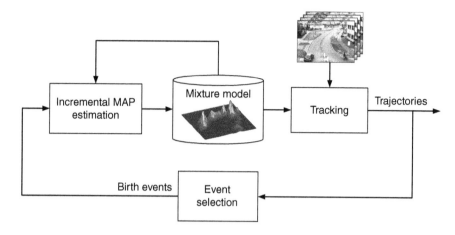

Figure 8.4 Learning strategy used to estimate the probability distribution of birth events.

- To remove trajectories generated by localised persistent *clutter*, we consider only objects that exhibit an average speed of at least a few pixels per second (e.g. 2 for a surveillance scenario with image sizes of 720×480 pixels) and whose product between the trajectory time span and speed is not negligible (e.g. larger than 10 pixels for a surveillance scenario with image sizes of 720×480 pixels). This last condition accounts for the fact that slower targets remain visible for longer. Note that the *thresholds* are defined in terms of seconds to provide independence from the frame rate.

Figure 8.5(a) shows the centroids of the birth event states obtained from the analysis of a 20-minute surveillance video clip. Clusters are mainly localised on the road and on the pavements. Note that birth events are also generated in non-entry areas because of track reinitialisations due to object proximity and occlusions.

To estimate $p(x|b)$, one can use a *semi-parametric* approach.[12] We approximate the distribution with mixture-of-Gaussian components that can be expressed as

$$p(x|b) \approx p(x|\vartheta) = \sum_{m=1}^{M} \pi_m p_m(x|\vartheta_m), \quad \text{with} \sum_{m=1}^{M} \pi_m = 1, \qquad (8.3)$$

---

[12]The advantage of using a semi-parametric approach is the compactness of the representation, compared with non-parametric approaches, as we do not need to store the original data used to compute the parameters.

(a)   (b)

Figure 8.5 The violet and blue dots superimposed on a sample image from the Broadway church scenario (BW) (CLEAR-2007 dataset) show the locations where target birth events (a) and clutter events (b) happened. IEEE © [1].

where

$$\vartheta = \{\pi_1, \ldots, \pi_M, \vartheta_1, \ldots, \vartheta_M\}$$

is the set of parameters defining the mixture. M is the number of components and

$$p_m(x|\vartheta_m) = \mathcal{N}(x, \mu_m, \Sigma_m)$$

is the mth Gaussian component with parameters

$$\vartheta_m = \{\mu_m, \Sigma_m\}.$$

μ_m and Σ_m are the mean and covariance matrix of the mth Gaussian, respectively.

The goal is to find the optimal set ϑ_{MAP} that maximises the log-posterior as

$$\vartheta_{\mathrm{MAP}} = \arg\max_{\vartheta} \left\{ \log p(X_b|\vartheta) + \log p(\vartheta) \right\}. \tag{8.4}$$

Starting with a large number of components, the algorithm converges towards the maximum a posteriori (MAP) estimate for ϑ by selecting the number of components necessary for the estimation using the Dirichlet prior

$$p(\vartheta) \propto \prod_{m-1}^{M} \pi_m^{-\tau}, \tag{8.5}$$

where $\tau = N/2$ and N is the number of parameters per component in the mixture.

In the Dirichlet distribution, τ represents the prior evidence of a component. When τ is negative (i.e. improper Dirichlet) the prior allows for the existence of a component only if enough evidence is gathered from the data. The prior drives the irrelevant components to extinction, thus favouring simpler models.

Although a good MAP solution can be reached by using the Expectation Maximisation (EM) algorithm [7], an online recursive procedure is preferable [8], as it allows one to refine the estimate with low computational cost once additional data become available.

Given the MAP estimate $\vartheta^{(n)}$ obtained using n data points

$$\{x^{(1)}, \ldots, x^{(n)}\}$$

and the new data $x^{(n+1)}$, we obtain the updated estimate $\vartheta^{(n+1)}$ by first computing the ownerships

$$o_m^{(n)}(x^{(n+1)}) = \pi_m^{(n)} p_m(x^{(n+1)}|\vartheta_m^{(n)})/p(x^{(n+1)}|\vartheta^{(n)}) \tag{8.6}$$

and by then updating the parameters as

$$\pi_m^{(n+1)} = \pi_m^{(n)} + \alpha \left(\frac{o_m^{(n)}(x^{(n+1)})}{1 - M\tau\alpha} - \pi_m^{(n)} - \frac{\tau\alpha}{1 - M\tau\alpha} \right), \tag{8.7}$$

for a Gaussian mixture with

$$p_m(x|\vartheta_m) = N(x, \mu_m, \Sigma_m)$$

then

$$\mu_m^{(n+1)} = \mu_m^{(n)} + \alpha \frac{o_m^{(n)}(x^{(n+1)})}{\pi_m^{(n)}} \left(x^{(n+1)} - \mu_m^{(n)} \right), \tag{8.8}$$

$$\Sigma_m^{(n+1)} = \Sigma_m^{(n)} + \alpha \frac{o_m^{(n)}(x^{(n+1)})}{\pi_m^{(n)}} \cdot$$

$$\cdot \left((x^{(n+1)} - \mu_m^{(n)})(x^{(n+1)} - \mu_m^{(n)})^T - \Sigma_m^{(n)} \right), \tag{8.9}$$

where α determines the influence of the new sample on the old estimate. A component m is discarded when the weight ϑ_m becomes negative.

Although mixture-of-Gaussian components may produce a good approximation of the underlying distribution using a small number of parameters, the final result may not be appropriate for tracking. In fact, *at initialisation*, when no prior information is available (i.e. $n = 0$), it is difficult to obtain a uniform distribution by mixing Gaussian components only.

As a solution, either we initialise Σ_m with large values, or we distribute a large number of components in the data space. Both approaches result

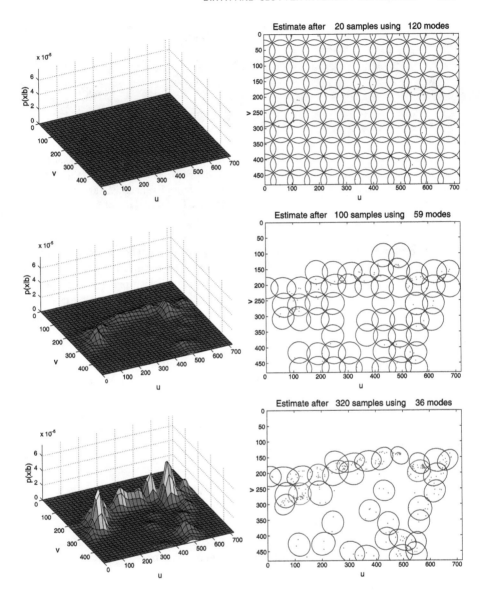

Figure 8.6 Example of online learning of the birth density $p(x|b)$ for the Broadway church scenario (BW), CLEAR-2007 dataset. The birth events used for density estimation are displayed in Figure 8.5(a). Although $p(x|b)$ is defined in a 6D state space, for visualisation purpose we show the information related to the 2D position subspace only (u, v). Left column: evolution of the birth density model with the number of samples processed. Right column: corresponding evolution of the GMM components. IEEE © [1].

Figure 8.7 Typical scenario where a dynamic occlusion may require track reinitialisation far from the typical target entry and exit areas. The white van at times occludes more than 50% of the view of the pedestrians. Eventually, target detection failures may cause lost tracks. Images from the PETS-2001 dataset.

in a slower learning process. Moreover, after training, the probability of an event far from the centre of the Gaussian tends to zero. If a birth or clutter event happens in these regions, then the tracking algorithm is likely to fail. A typical example of this problem is given by birth events generated by dynamic occlusions such as the ones shown in Figure 8.7. In this case, to avoid a lost track, a birth should be possible and the filter should include a delay before validating the object again.

To overcome this problem, we use a *non-homogeneous mixture* composed of a uniform component, $u(x)$, and the GMM of Eq. (8.3). We approximate $p(x|b)$ with

$$p(x|b) \approx \pi_u u(x) + \pi_g p(x|\vartheta), \tag{8.10}$$

where

$$u(x) = \frac{1}{V} rect(x),$$

V is the volume of the space, and π_u and π_g are the weights associated with the uniform component and with the Gaussian mixture.

We set at initialisation:

- $\pi_u = 1$ and $\pi_g = 0$ so that we have an uninformative initial estimate

- $\pi_u = 10^{-3}$ as the minimum value that π_u can get during learning.

Given this constraint, the algorithm refines $p(x|b)$ in a hierarchical fashion as follows:

- Use maximum likelihood (ML) to compute π_u and π_g (i.e. Eq. (8.6) and (8.7) with $\tau = 0$).

- Update ϑ independently from π_u according to Eqs. (8.6)-(8.9).

This approach introduces a bias in the estimate of the weights as the ownerships of Eq. (8.6) are computed using π_m and not $\pi_m \times \pi_g$. However, the update step of π_m and Σ_m does not depend on π_m (i.e. π_m simplifies by substituting Eq. (8.6) into Eq. (8.8) and Eq. (8.9)), and in practice with localised distributions and large n $\pi_u \ll \pi_g$, thus the bias tends to reduce with the amount of data available.

To *learn the birth intensity* from X_b, we initialise a grid of 12×10 6D Gaussians equally spaced in the 2D position subspace and centred on zero speed and on the objects' average size. The choice of the number of Gaussians depends on the complexity of the scene. However, as the components are selected by the Dirichlet prior (Eq. 8.5), we only need *to overestimate the number of entry regions*.

Figure 8.6 shows the evolution of the learning of $p(x|b)$ when additional data become available. The Dirichlet prior reduces the weight of the modes that are not supported by sufficient evidence. After processing 320 trajectories (Figure 8.6, last row, and Figure 8.8(a)) the peaks modelling the entry regions are clearly visible. Two major peaks correspond to areas over the road where vehicles appear. Smaller peaks are visible on the pavements. The remaining components of the mixture model birth events caused by track reinitialisations.

8.3.2 Clutter density

The procedure for the estimation of the clutter intensity $p_k(z|c)$ for Eq. (8.2) is similar to that of the birth intensity described in the previous section. However, unlike the estimation of birth intensity, the collection of the detections Z_c

(a)   (b)

Figure 8.8 Learned intensities for the Broadway church scenario (BW) from the CLEAR-2007 dataset superimposed on the original images. (a) Birth intensity (note that the major modes are associated with entry areas). (b) Clutter intensity (note that waving vegetation produces clutter that is correctly modelled by the GMM). IEEE © [1].

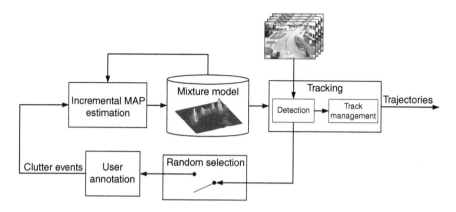

Figure 8.9   Learning strategy used to estimate the probability distribution of clutter events.

due to clutter is not performed using the tracker output, as short trajectories could also be generated by tracking errors on real objects.

Figure 8.10 shows an example of a flickering detection (and trajectory) on a small target. We use an *interactive approach* (see Figure 8.9) to select, in randomly displayed frames, the detections that are not associated with objects of interest. Figure 8.11 shows sample detections marked as clutter and Figure 8.5(b) displays the centroids of the clutter data collected on a real-world surveillance scenario. Most of the clutter is associated with waving vegetation. A few false detections are also associated with high-contrast regions due to shadows.

Given the sets of events

$$Z_c = \{z_{c,i}\}_{i=1}^{M_c},$$

representing the locations and sizes of the cluttered observations, we learn the clutter intensity by initialising a grid of 16×14 4D Gaussians equally spaced

Figure 8.10   Example of inconsistent detections interpreted by a PHD filter as clutter and therefore removed. Simple heuristics cannot differentiate these data from real clutter (red: observations; green: output of the PHD filter). IEEE © [1].

 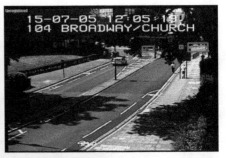

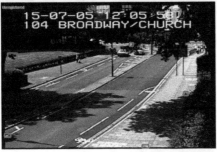

Figure 8.11 Sample detections that are marked interactively as clutter and then used for density estimation. IEEE © [1].

in the 2D position subspace and centred on the objects' average size. Because clutter data can be concentrated around small volumes of the observation space, we use a larger number of Gaussians than in the birth case to allow for higher spatial resolution.

Figure 8.12 and Figure 8.8(b) show an example of clutter density learned using 1800 false detections collected with user interaction on the results from the scenario Broadway church scenario (BW) (CLEAR-2007 dataset). The peaks of the probability distributions (in violet) correspond to areas of the image where waving vegetation generates a large number of false detections.

8.3.3 Tracking with contextual feedback

Figure 8.13 shows a block diagram of the overall tracking framework with context modelling.

Figure 8.14 shows sample results where contextual feedback improves the PHD filter performance. In this same scenario high birth intensity (i.e. weak filtering effect) is applied to the entry regions. This allows for correct detection and tracking of a fast car in the camera far-field. Similar considerations are valid for the clutter model. When clutter is localised, the GMM-based density estimation introduces further degrees of freedom for filter tuning. Further results and discussion are available in Section A.5 in the Appendix.

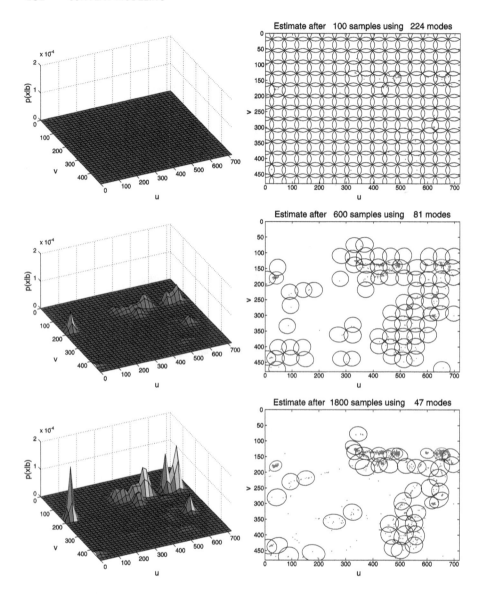

Figure 8.12 Example of online learning of the clutter density $p(x|c)$ for the Broadway church scenario (BW) from the CLEAR-2007 dataset. The input clutter events used are displayed in Figure 8.5(b). Although $p(x|c)$ is defined on a 4D observation space, for visualisation purpose we show the information related to the 2D position subspace only (u, v). Left column: evolution of the clutter density model with the number of samples processed. Right column: corresponding evolution of the GMM components. IEEE © [1].

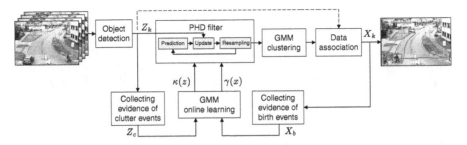

Figure 8.13 Multiple-target tracking scheme based on object detection and on particle PHD filtering. The PHD filter removes spatio-temporal noise from the observations before the tracker performs data association. Also, the output (Z_k) gathered by the detector and by the tracker (X_k) is used to extract statistics on the detector failure modalities $(\kappa(z))$ and on the object entry areas in the scene $(\gamma(x))$. The PHD filter uses this feedback to improve the tracking performance.

Figure 8.14 Filtering results of the particle-PHD filter using learned clutter and birth intensities (GM) on the same data as Figure 8.3(c) and (d). Top row: tracker output. Bottom row: the detections from a background subtraction algorithm are colour-coded in red and the PHD output is colour-coded in green. The filtering strength is modulated by the Gaussian mixture birth and clutter models. Weak filtering near an entry zone allows tracking the black car in the far-field.

8.4 SUMMARY

This chapter discussed the importance of context modelling for tracking and demonstrated how to learn contextual information using a combination of automated and interactive feedback from the tracker and the detector. This feedback is used to improve the accuracy of the overall detection and tracking results.

The natural modelling capabilities of the Bayesian multi-target tracking framework are used to locate where objects are more likely to appear in the scene (*birth events*) and where the detector is expected to produce errors (*clutter events*). Birth and clutter data are used to learn incrementally two parametric models (GMM). The PHD filter recursion uses these models to modulate the filter response depending on the location of the candidate targets. The tracking performance improvement is due to two factors: (i) the definition of a space-dependent birth model that allows the algorithm to increase the filtering strength where targets are unlikely to appear in the scene and to reduce the latency on real new born targets and (ii) the definition of a clutter model that increases the filter strength in the presence of spatially localised clutter.

REFERENCES

1. E. Maggio and A. Cavallaro. Learning scene context for multiple object tracking. *IEEE Transactions on Image Processing*, 18(8):1873–1884, 2009.

2. B.N. Vo, S.R. Singh and A. Doucet. Sequential Monte Carlo implementation of the PHD filter for multi-target tracking. In *Proceedings of the International Conference on Information Fusion*, Vol. 2, Cairns, Australia, 2003, 792–799.

3. H. Sidenbladh and S. Wirkander. Tracking random sets of vehicles in terrain. In *Proceedings of the IEEE Workshop on Multi-Object Tracking*, Madison, USA, 2003, 98.

4. D.E. Clark, I.T. Ruiz, Y. Petillot and J. Bell. Particle PHD filter multiple target tracking in sonar images. *IEEE Transactions on Aerospace and Electronic Systems*, 43(1):409–416, 2006.

5. N. Ikoma, T. Uchino and H. Maeda. Tracking of feature points in image sequence by SMC implementation of PHD filter. In *Proceedings of the SICE Annual Conference*, Vol. 2, Sapporo, Japan, 2004, 1696–1701.

6. Y.D. Wang, J.K. Wu, A.A. Kassim and W.M. Huang. Tracking a variable number of human groups in video using probability hypothesis density. In *Proceedings of the IEEE International Conference on Pattern Recognition*, Vol. 3, Hong Kong, China, 2006, 1127–1130.

7. M.A.T. Figueiredo and A.K. Jain. Unsupervised learning of finite mixture models. *IEEE Transactions on Pattern Analysis and Machine Intelligence*, 24(3):381–396, 2002.

8. Z. Zivkovic and F. Van der Heijden. Recursive unsupervised learning of finite mixture models. *IEEE Transactions on Pattern Analysis and Machine Intelligence*, 26(5):651–656, 2004.

9

PERFORMANCE EVALUATION

9.1 INTRODUCTION

The quality of a tracker depends on the task at hand and on the video content itself. No single tracker technique is universally useful for all applications. Moreover different trackers are not equally suited for a particular application.

Performance evaluation of video trackers involves comparing different tracking algorithms on a database of image sequences that includes representative application-related scenarios. Common practices for evaluating tracking results are based on human intuition or judgement (*visual evaluation*). To avoid the subjectivity of visual evaluation, an automatic procedure is desired. This procedure is referred to as *objective evaluation*. Objective evaluation measures have attracted considerable interest [1–7] and are used for the final assessment of an algorithm as well as for the development process to compare different parameter sets of an algorithm on large datasets.

An effective evaluation of a tracker is important for selecting the most appropriate technique for a specific application, and furthermore for optimally setting its parameters. However, except for well-constrained situations, the design of an *evaluation protocol* is a difficult task.

Video Tracking: Theory and Practice. Emilio Maggio and Andrea Cavallaro
© 2011 John Wiley & Sons, Ltd

A performance evaluation protocol is a pre-defined procedural method for the implementation of tracking experiments. It defines an evaluation method to enable the comparison of the results obtained with different video trackers. The main items to define when designing an evaluation protocol are the performance-evaluation scores, the dataset and, optionally, a ground truth:

- The performance evaluation *scores* quantify the deviation from the desired result, the computational resources that are necessary and other factors, such as, for example, latency in the generation of the tracking results and the number of track losses.

- The definition of an appropriate *dataset* for video tracking is challenging as the data must have enough variability to cover the application of interest so that the quality of the results on new data can be predictable. The recording (and the distribution) of a dataset is expensive and time consuming. Initially, the lack and cost of storage and computational resources limited the collection, sharing and use of sufficiently large datasets. More recently, privacy concerns have posed a barrier to lawful recording and distribution of real-world scenarios.

- A *ground truth* is the desired tracking result, generated manually or by the use of a gold-standard method. The generation of a ground truth is a time-consuming and error-prone procedure. The cost associated with the manual annotation of the dataset has limited the amount of data that can be used for objective comparisons of algorithms.

These three elements of a performance-evaluation protocol and objective evaluation strategies will be discussed in detail in the following sections.

9.2 ANALYTICAL VERSUS EMPIRICAL METHODS

The objective evaluation of a tracker may focus on judging either the tracking algorithm itself or the tracking results. These approaches are referred to as analytical methods or empirical methods, respectively.

Analytical methods evaluate trackers by considering their principles, their requirements and their complexity. The advantage of these methods is that an evaluation is obtained without implementing the algorithms. However, because trackers may be complex systems composed of several components, not all the properties (and therefore strengths) of a tracker may be easily evaluated.

Empirical methods, on the other hand, do not evaluate tracking algorithms directly, but indirectly through their results. To choose a tracker based on empirical evaluation, several algorithms (or parameter sets for the same algorithm) are applied on a set of test data that are relevant to a given application.

The algorithm (or parameter set) producing the best results should be then selected for use in that application.

Empirical methods can be divided into standalone and discrepancy methods. Empirical *standalone* methods do not require the use of a ground truth [8, 9]. These methods measure the 'goodness' of a video-tracking result based on some quality criteria that can be defined about a trajectory. Examples of quality criteria can be related to the smoothness of the trajectory or to the verification of the consistency of results with time-reversed tracking [8].

Empirical *discrepancy* methods quantify the deviation of the actual result of the tracker from the ideal result. These methods aim to measure the tracking result compared to a reference result, the ground truth. The measures quantify spatial as well as temporal deviations. Empirical discrepancy methods can be classified into two groups, namely low-level and high-level methods:

- Low-level methods evaluate the results of a tracker per se, independently from an application. For example, Video Analysis and Content Extraction (VACE) is a low-level empirical evaluation protocol (see Section 9.6.1.2).

- High-level methods evaluate the results of a tracker in the context of its application. For example, Imagery Library for Intelligent Detection Systems (i-Lids) and Challenge of Real-time Event Detection Solutions (CREDS) are high-level empirical evaluation protocols (see Section 9.6.2.3 and Section 9.6.2.1, respectively).

The Evaluation du Traitement et de l'Interpretation de Sequences vidEO (ETISEO) combines both low-level and high-level evaluation (see Section 9.6.2.2).

A summary of performance-evaluation strategies for video-tracking algorithms is shown in Figure 9.1.

9.3 GROUND TRUTH

The ground truth represents the desired (ideal) tracking result, generated either manually or by the use of a gold-standard tracking method. A ground-truth can refer to different properties of the accuracy of the tracking results:

- The presence of the target projection on the image in a particular *pixel* at a particular time (pixel-level ground truth)

- The value of the target *state* or subset of the state parameters at a particular time (state-level ground truth)

- The time interval corresponding to the *life span* of the target.

The ground truth at pixel level is often approximated with a bounding box or ellipse around the target area (see Figure 9.2). The ground-truth definition

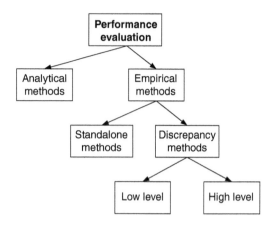

Figure 9.1   Taxonomy of performance-evaluation strategies for video-tracking algorithms.

rules at pixel level generally also define how to treat partially non-visible objects in the following situations:

- A target appearing in or disappearing from the field of view
- A target undergoing a partial or a total occlusion.

For example, the ground truth could be generated (or used in the evaluation of the performance scores) only for frames when a target appears fully in the field of view of the camera. In Section 9.7 we provide examples of datasets with available ground-truth annotation.

Generating the ground truth at pixel and state level is considerably more time consuming than generating it for the life span of the targets. To simplify the ground-truth creation process at pixel (or at state level), a common approach is to generate it on a subset of the total number of frames only, or to define regions of no interest (*no-care areas*) where results are not expected and therefore will not be annotated. Examples of no-care areas are portions of the far-field of the camera (where objects are too small) or regions in the image that are occluded by vegetation.

As the generation of ground-truth data is error prone, it is important to assess the *accuracy of the ground-truth* data itself and, in the case of manual annotation, the precision across different annotators. This information must be used to assess the confidence level of the performance evaluation (i.e. of the evaluation scores) based on this ground-truth data.

When *multiple targets* are simultaneously present in the scene, the target labels in the ground truth do not necessarily correspond to the target labels in the tracker estimation results. Therefore, unless a tracker has also recognition capabilities (i.e. the ability to associate a unique pre-defined identity to each object), an association step between ground truth and estimates is necessary.

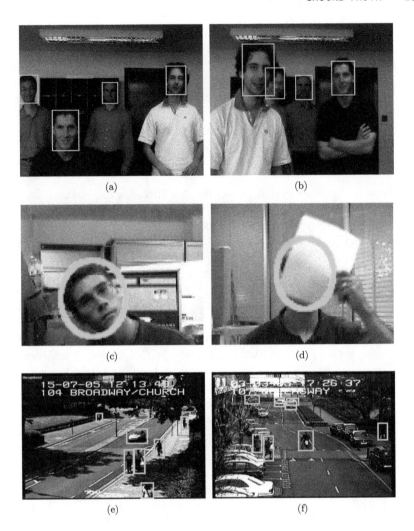

Figure 9.2 Example of ground-truth annotation at pixel level on standard test sequences. Note the different approximation of the target areas (bounding box or bounding ellipse) and the definition of the annotation in case of partial and total occlusions. Images from: (a) and (b) SPEVI dataset; (c) and (d) Birchfield head-tracking dataset; (e) and (f) i-Lids (CLEAR-2007) dataset reproduced with permission of HOSDB.

Let $B = \{\mathbf{b}_1, \ldots, \mathbf{b}_N\}$ be the set output of the tracking algorithm, where $\mathbf{b}_j$ is the output for the jth target, and $\tilde{B} = \{\tilde{\mathbf{b}}_1, \ldots, \tilde{\mathbf{b}}_{\tilde{N}}\}$ be the ground-truth set. The elements of these sets can be:

- target bounding boxes at a frame k
- target trajectories
- target life spans.

The *association* between ground truth and estimated data can be formulated as an assignment problem. To solve this assignment problem methods such as nearest neighbour and graph matching can be used, as discussed in Section 7.3. Once the association is performed, the multi-target evaluation can be cast onto multiple single-target evaluations using the scores discussed in the next section.

9.4 EVALUATION SCORES

This section discusses evaluation scores that measure the quality of tracking results. Different scores highlight different properties of the trajectory-estimation errors. In particular, trajectory estimation can be seen either as a localisation problem or as a classification problem. We will introduce the evaluation scores for both scenarios and discuss their applicability.

9.4.1 Localisation scores

A straightforward solution to objectively compare different video-tracking results is to compute the distance between the state estimate and the ground-truth state for each frame. Particular care has to be taken when comparing errors representing areas and when evaluating multiple-target trackers.

9.4.1.1 *Single-target scores*   A simple measure of the error between the state estimate x and the ground-truth state $\tilde{x}$ is the Euclidean distance, $d_2(.)$, defined as

$$d_2(x, \tilde{x}) = \sqrt{(x - \tilde{x})'(x - \tilde{x})} = \sqrt{\sum_{i=1}^{D}(x_i - \tilde{x}_i)^2}, \qquad (9.1)$$

where D is the dimension of the state space E_s.

When it is desirable to assign a higher weight to large errors, one can substitute the 2-norm with a p-norm, with $p > 2$, defined as

$$d_p(x, \tilde{x}) = \sqrt[p]{\sum_{i=1}^{D} \|x_i - \tilde{x}_i\|^p}. \qquad (9.2)$$

The *overall trajectory error*, $e_p(.)$, can be obtained by averaging the result over the K frames of the life span of a target as

$$e_p(x_{1:K}, \tilde{x}_{1:K}) = \frac{1}{K}\sum_{k=1}^{K} d_p(x_k, \tilde{x}_k). \qquad (9.3)$$

When it is desirable to *highlight the importance of different state parameters* (for example by weighting more the contribution of positional error with respect to the contribution of shape and velocity errors), one can replace the Euclidean distance with the *Mahalanobis distance*, $d^{(M)}(.)$, defined as

$$d^{(M)}(x, \tilde{x}) = \sqrt{(x - \tilde{x})'\Sigma^{-1}(x - \tilde{x})}, \tag{9.4}$$

where Σ can be, for example, a diagonal covariance matrix whose diagonal elements reweight the contribution of each state parameter.

9.4.1.2 Dealing with the perspective bias

A problem arises when using distance metrics on positional state estimates like the ones produced by elliptic and box-based trackers (see Section 1.3.1). As the centroid error score does not depend on the target size, an estimation error of a few pixels for an object close to the camera counts the same as the estimation error for the same object when positioned in the far-field of the camera. However, when an object is far from the camera its projection on the image plane is relatively smaller than when it is closer. Therefore, small positional errors in the far-field may correspond to track estimates that do not share any pixels with the ground truth. To solve this problem, one can compute a centroid error *normalised by the target size*.

As an example, let the state x of an elliptical tracker

$$x = (u, v, w, h, \theta)$$

be represented by the centroid (u, v), the length of the two semi-axes (w, h) and the rotation of the ellipse θ. Also, let us assume that ground-truth information

$$\tilde{x} = (\tilde{u}, \tilde{v}, \tilde{w}, \tilde{h}, \tilde{\theta})$$

on the expected position of the ellipse is available. The centroid error can be normalised by rescaling and rotating the coordinate system so that the errors in u and v become

$$e_u(x, \tilde{x}) = \frac{\cos\tilde{\theta}(u - \tilde{u}) - \sin\tilde{\theta}(v - \tilde{v})}{\tilde{w}}, \tag{9.5}$$

$$e_v(x, \tilde{x}) = \frac{\sin\tilde{\theta}(u - \tilde{u}) + \cos\tilde{\theta}(v - \tilde{v})}{\tilde{h}}. \tag{9.6}$$

From Eqs. (9.5) and (9.6) follows the definition of a normalised Euclidean error at frame k as

$$e(x_k, \tilde{x}_k) = \sqrt{e_u(x_k, \tilde{x}_k)^2 + e_v(x_k, \tilde{x}_k)^2}. \tag{9.7}$$

Note that, when the estimated centroid is outside the ground-truth area, then $e(x, \tilde{x}) > 1$.

9.4.1.3 Multi-target scores The error scores defined so far are applicable to single targets. As discussed in Section 9.3, when the ground-truth data describe the trajectory of multiple targets, an *association* between ground-truth trajectories and the tracker-output trajectories is required before evaluating the performance.

A comprehensive solution to extend the Euclidean distance to multiple targets and to include the association step directly in the evaluation is provided by the Wasserstein distance [10]. The Wasserstein distance is an evaluation metric for multi-target algorithms and has been used to evaluate trackers based on the RFS framework [11–13].

Let $f(x)$ and $g(x)$ be two probability densities over the Euclidean state space E_s and $h(x, y)$ a joint distribution with marginals $f(x)$ and $g(y)$. Then we can define the Wasserstein distance, $d_p^{W}(.)$, between the two distributions f and g as

$$d_p^{W}(f, g) = \inf_h \sqrt[p]{\int d_p(x, y)\, h(x, y) \mathrm{d}x \mathrm{d}y}. \tag{9.8}$$

The multi-target interpretation of the distance is obtained by formulating Eq. (9.8) for two finite sets

$$X_k = \{x_{k,1}, \ldots, x_{k,n}\}$$

and

$$Y_k = \{y_{k,1}, \ldots, y_{k,m}\}.$$

These two finite sets represent two possible multi-target results composed of a finite collection of single-target states. To this end we define the distribution over the state space E_s of a finite set X_k as

$$\delta_{X_k}(x) = \frac{1}{n} \sum_{i=1}^{n} \delta(x_{k,i}),$$

where the $\delta(.)$ are Dirac's deltas centred on the elements of the multi-target finite set.

By substituting δ_{X_k} and δ_{Y_k} into f and g, the result is a metric on the multi-target space of the finite subsets. The result of the substitution is

$$d_p^{W}(X, Y) = d_p^{W}(\delta_X, \delta_Y) = \inf_C \sqrt[p]{\sum_{i=1}^{n} \sum_{j=1}^{m} C_{i,j} d_p(x_i, y_j)}, \tag{9.9}$$

where, for simplicity of notation, we dropped the frame index k.

C is the $n \times m$ transportation matrix to be optimised under the constraints

$$\forall i,j \ \ C_{i,j} > 0,$$

$$\forall j \ \sum_{i=1}^{n} C_{i,j} = \frac{1}{m},$$

$$\forall i \ \sum_{j=1}^{m} C_{i,j} = \frac{1}{n}.$$

The computation of the Wasserstein distance of Eq. (9.9), i.e. the optimisation of the matrix C, can be formulated as a linear assignment problem similar to those described in Section 7.3.2. The optimal solution to this problem can be found using the algorithm from Hopcroft and Karp [14]. We briefly summarise here the steps necessary to obtain the new formulation [10].

Let us define c as the greatest common divisor of m and n, and let

$$\hat{n} = m/c,$$

$$\hat{m} = n/c.$$

The size $N = n\hat{n} = m\hat{m}$ assignment problem is between the finite set $\hat{X}$ formed by substituting in X the vectors x_i with $\hat{n}$ arbitrarily close replicas $x_{i,1}, \ldots, x_{i,\hat{n}}$ and the similarly composed finite set $\hat{Y}$. From Eq. (9.9) we can write

$$d_p^{\mathrm{W}}(X,Y) = d_p^{\mathrm{W}}(\hat{X},\hat{Y}) = \inf_{\nu} \sqrt[p]{\sum_{i=1}^{n} \sum_{\hat{i}=1}^{\hat{n}} d_p(\hat{x}_{i,\hat{i}}, \hat{y}_{\nu(i,\hat{i})})}, \qquad (9.10)$$

where ν defines a possible association between the N pairs $(i, \hat{i})$ and the N pairs $(j, \hat{j})$.

9.4.2 Classification scores

Another approach to evaluating a tracker is to assess its performance in a classification problem. This classification problem may compare the estimation of the tracker with the ground-truth data at different levels, namely pixel, object, trajectory or temporal (see Figure 9.3):

- At *pixel* level, one evaluates whether the pixels belong to the estimated target area only, to the ground-truth area only or to both.

- At *object* level, one evaluates the correspondences between ground-truth targets and estimated targets, considering the target as a whole.

- At *trajectory* level, one evaluates the correspondence between ground-truth and estimated trajectories.

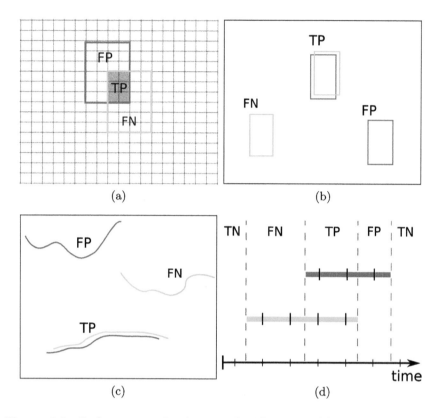

Figure 9.3 Performance evaluation as a classification problem. The green items represent the ground-truth data, while the blue items represent the estimates (output) of the tracker. True positives (TP), false positives (FP) and false negatives (FN) at (a) pixel level, (b) object level, (c) trajectory level and (d) temporal level.

- At *temporal* level, one classifies the correspondence between the real and the estimated target life span.

When considering performance evaluation as a classification problem, the tracking results can be assessed based on true positive (TP), false positive (FP), false negative (FN) and true negative (TN) classifications:

- The number of TPs is the number of successful classifications. For example, at pixel level, the pixels within the estimated target area overlapping the ground-truth are true positives (see Figure 9.3(a)).

- An FP is declared when an hypothesis of target presence (at pixel, object, trajectory or temporal level) is incorrect. For example, at object level, a false positive is declared when the estimate does not *significantly* overlap any objects in the ground truth (see Figure 9.3(b)).

- The number of FNs is obtained by counting the number of ground-truth items (pixels, objects, trajectories or frames) that are not matched by any tracker estimates. For example, at temporal level, a false negative is a frame when a track is lost.

- A TN is awarded when the estimation of the non-existence of a target is correct. Note that the number of TNs can be meaningfully defined only at the pixel and temporal level. Moreover, if in the ground truth the number of negative items (e.g. background pixels and frames without a target) far exceeds the number of positive items, then the the number of TNs shall be disregarded in the performance evaluation.

Based on the above definitions, we can define as performance evaluation scores the precision, the recall, the F-score and the specificity.

The *precision*,[13] $\mathcal{P}$, is the fraction of correct results with respect to all the generated estimation results

$$\mathcal{P} = \frac{|\text{TP}|}{|\text{TP}| + |\text{FP}|}, \tag{9.11}$$

where $|.|$ denotes the cardinality of a set.

The *recall*, $\mathcal{R}$, is the fraction of correct results with respect to the expected results

$$\mathcal{R} = \frac{|\text{TP}|}{|\text{TP}| + |\text{FN}|}, \tag{9.12}$$

The *F-score*, $\mathcal{F}$, is the harmonic mean of precision and recall

$$\mathcal{F} = 2\frac{\mathcal{P} \cdot \mathcal{R}}{\mathcal{P} + \mathcal{R}}. \tag{9.13}$$

Its weighted version, $\mathcal{F}_\beta$, is defined as

$$\mathcal{F}_\beta = (1 + \beta^2)\frac{\mathcal{P} \cdot \mathcal{R}}{\beta^2 \cdot \mathcal{P} + \mathcal{R}}, \tag{9.14}$$

where the weight β modulates the importance of the precision with respect to the recall.

Finally, when the use of TN is meaningful, one can evaluate the *specificity*, $\mathcal{S}$, of an algorithm as

$$\mathcal{S} = \frac{|\text{TN}|}{|\text{TN}| + |\text{FP}|}. \tag{9.15}$$

---

[13] The meaning of the term precision here refers to a quantity evaluated over a classification result. Its meaning is different from the meaning of the term precision defined in Chapter 1, which refers instead to the degree of reproducibility of a tracking estimation.

It is important to notice that several variations in the score in Eq. (9.11) can be used. A measure of *overlap* $O^{(1)}$ between the tracker estimation and the ground truth, also known as dice, is given by

$$O^{(1)} = \frac{2 \cdot |TP|}{2 \cdot |TP| + |FP| + |FN|}, \tag{9.16}$$

The dice error, $\mathcal{D}$, is then defined as

$$\mathcal{D} = 1 - O^{(1)}; \tag{9.17}$$

and we will denote the dice error in a single frame as $\mathcal{D}_k$.

Other measures of overlap are the following:

$$O^{(2)} = \frac{|TP|}{|TP| + |FP| + |FN|}, \tag{9.18}$$

and

$$O^{(3)} = \frac{|TP|^2}{2 \cdot |TP| + |FP| + |FN|}, \tag{9.19}$$

These three overlap scores, $O^{(1)}$, $O^{(2)}$ and $O^{(3)}$, are bounded between 0 and 1 and reward tracking estimates with a high percentage of true positive pixels, and with few false positives and false negatives. However, as the intersection of the target and ground truth areas decreases, the score $O^{(2)}$ decreases faster than $O^{(1)}$.

It is interesting to notice that unlike centroid-based measures, when applied at the pixel level, these scores can account for errors in the estimation of the *target size*. Moreover, in the case of a lost track, they do not depend on where in the image the tracker is attracted to (i.e. $\mathcal{D}$ saturates to 1).

9.5 COMPARING TRACKERS

A summary of the performance of a video tracker over selected test sequences can be given by the average score of the measures discussed in the previous section. However, these average scores should be complemented with additional information and one has to consider also:

- possible *lost tracks*
- the *significance* of the results
- the *precision* of the results,

as discussed in the next sections.

9.5.1   Target life-span

The performance scores presented in Section 9.4 compute a global estimate of tracking quality and do not explicitly consider the case of lost tracks. A track loss happens when the target is present, but the tracker estimation is 'far' from it. The meaning of 'far' depends on the application.

When tracking fails and a target is lost, unless the corresponding track is terminated, the tracking algorithm tends to *lock* on an area of the background with features that are similar to those of the lost target. In this situation, the value of a centroid-based error measure will depend on the location of this lock area. However, from the evaluation point of view the magnitude of the error is an irrelevant number and therefore should not be used in the evaluation of a score.

To associate a score with this process, one can calculate the *lost track ratio* λ, the ratio between the number of frames where the tracker is not successful N_s, and the total number of frames in a test sequence N_t:

$$\lambda = \frac{N_s}{N_t}. \tag{9.20}$$

The use of this score requires the definition of when to declare a track to be lost. For example, a track at time index k can be defined *lost* when the dice error (Eq. 9.17) in that frame $\mathcal{D}_k$ exceeds a certain value T_λ:

$$\mathcal{D}_k > T_\lambda. \tag{9.21}$$

Typically T_λ is in the range $[0.8, 1]$.

Particular care is then necessary when comparing other scores, as their values may be dependent on λ: a smaller λ might mean that the performance scores are *computed on a larger number of frames*. As the additional frames are usually the most challenging parts of a sequence (i.e. where some trackers failed), the value of λ has to be considered first in the evaluation. Then, only if the values of λ obtained by two trackers are comparable, can one compare the values of the other performance scores.

In summary, given the ground truth, the dataset and the selected scores, the evaluation procedure should be divided in the following steps:

1. Choose T_λ that is appropriate for the application of interest.

2. Identify the frames with lost tracks (based on Eq. 9.21) and remove them from the computation of the performance scores.

3. Compute λ for each algorithm under evaluation.

4. If two trackers have a similar value of λ, then compare their performance scores obtained in Step 2. Otherwise, the value of λ will suffice for the assessment of the performance.

9.5.2 Statistical significance

The comparison of the average scores can be insufficient to reach a meaningful conclusion on which of the algorithms under evaluation is preferable, as performance improvements may be not consistent over a whole dataset, composed of N test sequences. To overcome this problem, *statistical hypothesis testing*, like the Student's t-test [15], can be used.

The t-distribution is the distribution of the mean of a normally distributed small number of samples. Under the assumption that two algorithms with identical performance produce normally distributed score differences, one can test the hypothesis that the mean difference is significant against the hypothesis that the difference is due to random fluctuations. To perform the test, one first computes the $\hat{t}$ value as

$$\hat{t} = \frac{E[e_a - e_b]}{\sigma_d/\sqrt{N}}, \tag{9.22}$$

where $E[e_a - e_b]$ is the mean score difference between two algorithms a and b, σ_d is the standard deviation of the sample differences, and N is the number of samples (i.e. in our case the number of sequences in the dataset).

The probability that this particular value of $\hat{t}$ occurred by chance is given by the *p*-value. The *p*-value is the value of the cumulative function of the Student t-distribution evaluated in $\hat{t}$. If the *p*-value is lower than a pre-defined threshold of statistical significance (e.g. 0.05 or 0.01), then the hypothesis that the difference between the mean scores is statistically significant is accepted.

9.5.3 Repeatability

Special attention is required when evaluating *non-deterministic* tracking methods, such as a particle filter. In fact, due to random sampling, the performance of the algorithm may not be consistent over different runs. It is therefore necessary to repeat the experiment several times under the same conditions and then to consider the average score over these runs, its standard deviation and, optionally, its maximal and minimal values.

A good tracker is characterised *not only by a small average error, but also by small variations in the error*. A measure of consistency is given by the covariance of the performance score.

A *repeatability coefficient* can be considered that represents the value below which the discrepancy between the performance score under analysis in two difference runs of the algorithm is expected to lie with a certain probability (e.g. 0.95).

Also, to test different parameter sets, one can apply again hypothesis testing of the unpaired (or grouped) variety [16], where in this case the samples are multiple runs over the same sequence.

9.6 EVALUATION PROTOCOLS

A performance-evaluation protocol is a pre-defined procedural method for the implementation of tracking experiments and their subsequent evaluation. Evaluation protocols[14] standardise evaluation methods to enable the comparison of the results generated by different video trackers. Evaluation protocols include in general test sequences, ground-truth data, evaluation rules and scoring tools.

9.6.1 Low-level protocols

This section covers two low-level evaluation protocols, namely:

- The Evaluation du Traitement et de l'Interpretation de Sequences vidEO (ETISEO) [17] protocol comparing target detection, localisation, and multi-target tracking.

- The Video Analysis and Content Extraction (VACE) protocol [18], used in the evaluation challenges that took place during the Classification of Events, Activities and Relationships (CLEAR) workshops [19] and the TREC Video Retrieval Evaluation campaigns [20].

9.6.1.1 ETISEO   The ETISEO protocol is based on the evaluation of the number of TPs, FPs, and FNs at object and trajectory level.

The *data-association* step between tracker estimations and ground-truth data is based on a simple nearest neighbour approach. The association hypotheses are weighted either by one of the overlap scores defined in Eq. (9.16), (9.18) and (9.19), or by a fourth score, $O^{(4)}$, defined as

$$O^{(4)} = \max\left\{ \frac{|\text{FP}|}{|\text{FP}| + |\text{TP}|}, \frac{|\text{FN}|}{|\text{FN}| + |\text{TP}|} \right\}. \tag{9.23}$$

In the following we outline the scores used in the ETISEO object-detection and trajectory-evaluation tasks. The *object-detection* evaluation task is divided into three parts:

1. The first part of the performance evaluation compares in each frame the estimated *number of targets* in the scene N against the ground-truth datum, $\tilde{N}$. Precision (Eq. 9.11) and recall (Eq. 9.12) are used, with the

---

[14]Note that the terminology that will be used in this section is that of the specific protocols and might differ from the formal definitions we gave elsewhere in the book.

following definitions for the quantities in Eq. (9.11) and Eq. (9.12) as

$$|\text{TP}| = \min\{N, \tilde{N}\}$$
$$|\text{FP}| = N - |\text{TP}|$$
$$|\text{FN}| = \tilde{N} - |\text{TP}|.$$

2. The second part requires *association* between the estimated detections and the ground-truth detections. Given the association, precision (Eq. 9.11) and recall (Eq. 9.12) scores are again computed. In this case, unpaired detections and unpaired ground-truth data are treated as FP and FN, respectively.

3. The third part evaluates the *spatial overlap* between ground-truth data and the estimated bounding boxes (or blobs). To this end, six scores are uses: precision, recall, specificity and F-score, computed at the pixel level (see Eq. (9.11)–(9.13) and Figure 9.3(a)), and two additional scores that measure the level of ambiguity of the spatial estimate (target split and target merge). The rationale for the use of these two additional scores is that a typical error in object-detection algorithms based on background subtraction and learned classifiers is to generate multiple detections (or blobs) for a single target. To measure the level of ambiguity one can first count the number of detections (or blobs) $\tilde{D}_i$ that can be assigned to the ith object in the ground-truth data. The assignment is computed by gating the data according to one of the overlap measures defined in Eqs. (9.16), (9.18) and (9.19) and a fixed threshold. From this assignment one can derive a *target split* score defined as

$$S = \frac{1}{K} \sum_{k=1}^{K} \frac{1}{\tilde{N}_k} \sum_{i=1}^{\tilde{N}_k} \frac{1}{\tilde{D}_i}. \tag{9.24}$$

A second score evaluates the situation of one detection (or blob) bounding the area of two ground-truth targets. By counting the number of targets in the ground-truth D_i that can be associated to the ith estimated bounding box (or blob), the *target merge* score M can be computed as

$$M = \frac{1}{K} \sum_{k=1}^{K} \frac{1}{N_k} \sum_{i=1}^{N_k} \frac{1}{D_i}. \tag{9.25}$$

The *trajectory management* evaluation task assesses the consistency of trajectory associations over time:

- For each pair of consecutive frames at time k and $k+1$, one can compare the temporal connections between the estimated detections and the connections in the ground-truth data. If the *link* connects two detections that are associated with the same ground-truth object, then this is considered as a TP. Missing links, and additional links are treated as FNs and FPs, respectively. From these definition and Eqs. (9.11), (9.12), and (9.13), one can derive the corresponding precision, recall and F-score values for links.

- The same criterion can be extended to multiple frames by assigning to the ground-truth data full trajectories instead of frame-to-frame links. A *trajectory* is declared a TP when the percentage of consistent single-frame associations is above a pre-defined threshold. The definition of FP and FN trajectories as well as the definition of tracking precision, sensitivity and F-score is then straightforward.

- In addition to these scores, ETISEO evaluates the *temporal overlap* between each of the $\tilde{N}$ ground-truth trajectories $\{\tilde{t}_1, \dots, \tilde{t}_{\tilde{N}}\}$ and the associated estimates $\{t_1, \dots, t_{\tilde{N}}\}$, where each t_i is either the first trajectory that spatially overlaps the ith ground-truth trajectory, or the trajectory t with maximum temporal overlap $|t \cap \tilde{t}_i|$. The temporal overlap, O, representing the average percentage of time the objects are detected and tracked is computed as

$$O = \frac{1}{\tilde{N}} \sum_{i=1}^{\tilde{N}} \frac{|t_i \cap \tilde{t}_i|}{|\tilde{t}_i|}. \tag{9.26}$$

To assess the problem of multiple estimates overlapping the trajectory of a single object, ETISEO defines the *persistence* score P as

$$P = \frac{1}{\tilde{N}} \sum_{i=1}^{\tilde{N}} \frac{1}{\tilde{D}_i^{(t)}}, \tag{9.27}$$

where $\tilde{D}_i$ is the number of estimated trajectories that, in at least one frame, are the best assignment for the ith ground-truth datum.

Similarly, to assess the problem of one trajectory estimate overlapping multiple ground-truth data, ETISEO defines the *confusion* score C as

$$C = \frac{1}{N} \sum_{i=1}^{N} \frac{1}{D_i^{(t)}}, \tag{9.28}$$

where $D_i^{(t)}$ is the number of ground-truth objects that at least in one frame are the best possible association for the ith estimate.

9.6.1.2 *VACE* The VACE protocol uses four main *evaluation scores*:

- two scores assess the accuracy of target localisation and area estimation, and

- two scores evaluate the overall tracking performance.

The accuracy of an estimated bounding box is quantified with the overlap measure $O^{(2)}$, computed at pixel level (see Eq. (9.18) and Figure 9.3(a)). Slack variables with value 0.2 are used in order to model missing detections and clutter. The results from the *data association* step at each frame k are:

- a set of $M_k^{(d)}$ matched bounding boxes pairs,[15] with overlap scores $O_{k,1}, \ldots, O_{k,M_k^{(d)}}$ and

- the count of ground-truth and estimated detections without any association, namely the number of false positives, $|\text{FP}_k^{(d)}|$, and the number of false negatives, $|\text{FN}_k^{(d)}|$.

The accuracy of *target localisation and area estimation* is quantified with the so-called multiple object detection precision (MODP), the average overlap over the matching data:

$$\text{MODP}(k) = \frac{1}{M_k^{(d)}} \sum_{i=1}^{M_k^{(d)}} O_{k,i}. \tag{9.29}$$

and the multiple object detection accuracy (MODA), which is related to the cost of false positives and false negatives:

$$\text{MODA}(k) = 1 - \frac{c_n\left(|\text{FN}_k^{(d)}|\right) + c_p\left(|\text{FP}_k^{(d)}|\right)}{|\tilde{B}_k|}, \tag{9.30}$$

where $c_p(.)$ and $c_n(.)$ are the cost functions for the false positives and false negatives respectively, and $|\tilde{B}_k|$ is the number of elements in the ground truth. The global scores for a video sequence are obtained by averaging MODP and MODA over the frames.

The *overall tracking performance* is obtained with the so-called multiple object tracking precision (MOTP):

$$\text{MOTP} = \frac{\sum_{i=1}^{M^{(t)}} \sum_{k=1}^{K} O_{k,i}}{\sum_{k=1}^{K} M_k}, \tag{9.31}$$

where M_k is the number of mapped objects over the entire test sequence; and the multiple object tracking accuracy (MOTA):

$$\text{MOTA} = 1 - \frac{\sum_{k=1}^{K} \left(c_n\left(|\text{FN}_k^{(t)}|\right) + c_p\left(|\text{FP}_k^{(t)}|\right) + \log_e\left(id_{\text{sw}}\right)\right)}{\sum_{k=1}^{K} |\tilde{B}_k|}, \tag{9.32}$$

where $\text{FN}_k^{(t)}$ and $\text{FP}_k^{(t)}$ denote the sets of missing and false-positive tracks, and id_{sw} is the number of identity switches. The cost functions $c_n(.)$ and $c_p(.)$

---

[15] The superscript d indicates a *detection*-level association (target area). Similarly, the superscript t will indicate a *tracking*-level association, when the data-association problem will be solved at the trajectory level (Figure 9.3(c)).

can be defined by the user to adjust value of the penalty associated with false positives and false negatives.

9.6.2 High-level protocols

As discussed in Section 9.1, the performance of a video tracker can be measured *indirectly* by analysing the performance of subsequent processing modules, such as event detection, making use of tracking estimations. We discuss below a series of evaluation protocols and a set of scores for event-detection tasks.

9.6.2.1 CREDS The evaluation procedure of the Challenge of Real-time Event Detection Solutions (CREDS) [21] is aimed at comparing event-detection algorithms for surveillance in underground transport networks (Figure 9.4).

The CREDS ground truth contains annotation for a set of nine classes of events (e.g. proximity to the platform edge, walking on the tracks). The events are further classified according to their priority as warnings, alarms and critical alarms. The goal is to trigger the appropriate alarm at the correct time and to estimate the duration of the event. Errors are weighted differently, depending on the event priority:

- TPs are awarded when ground-truth events overlap with automatically generated events of the same type.

- FNs are awarded for undetected events.

Figure 9.4 Sample frames from two scenarios of the CREDS dataset. Courtesy of RATP.

- False Positives are classified into two groups:

    - FP$_1$: events overlapping ground-truth data that are already assigned to other events.

    - FP$_2$: false detections that do not overlap with any data in the ground truth.

The performance of the algorithm is summarised by a global score S, which sums the scores assigned to TPs and penalties assigned to FPs and FNs, that is

$$S = |\text{FP}_1| \cdot P_{\text{FP}_1} + |\text{FP}_2| \cdot P_{\text{FP}_2} + |\text{FN}| \cdot P_{\text{FN}} + \sum_{i=0}^{|\text{TP}|} S_{\text{TP}}^{(i)},$$

where S_{TP} is a score that depends on the quality of each TP, while P_{FP_1}, P_{FP_2}, and P_{FN} are penalties that depend on the event priority (i.e. warning, alarm, or critical alarm).

The score S_{TP} depends on the precision in detecting the event and estimating its duration. Anticipated and delayed detections are scored according to a function of the time difference between estimate and ground truth Δt (Figure 9.5), defined as

$$S_{\text{TP}} = \begin{cases} 0 & \Delta t \in \ ]-\infty, B[\ \cup \ ]D, -\infty[\\ \dfrac{S_{max}}{A-B}(\Delta t - B) & \Delta t \in \ [B, A] \\ S_{max} & \Delta t \in \ ]A, 0[\\ \dfrac{S_{max}}{D}(D - \Delta t) & \Delta t \in \ ]0, D[\end{cases}.$$

The maximum value $S_{\max}$ is a parabolic function of ratio between event duration estimate d and actual duration $\tilde{d}$, as defined in the ground-truth

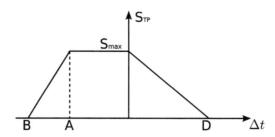

Figure 9.5   CREDS evaluation score as a function of the anticipation/delay time Δt between the automatically detected event and the ground-truth event. Anticipated detections are favoured over delayed detections.

Table 9.1 Numerical values of the parameters defining the CREDS evaluation score. The values depend on the event priority.

	A (ms)	B (ms)	D (ms)	P_{FN}	P_{FP_1}	P_{FP_2}
Warnings	-1000	-2000	2000	-100	-5	-50
Alarms	-1000	-2000	1000	-250	-5	-80
Critical Alarms	-1000	-2000	200	-1000	-5	-200

annotation, that is

$$
S_{\max} = \begin{cases} 50 \cdot \left[2 - \left(1 - d/\tilde{d} \right)^2 \right] & d/\tilde{d} \in [0,2] \\ 50 & d/\tilde{d} \in \,]2,\infty] \end{cases}.
$$

The score parameters A, B and D as well as the false positive and false negative penalties P_{FP_1}, P_{FP_2} and P_{FN} depend on the event priority and are summarised in Table 9.1. The values are chosen in such a way that the penalty for errors and delays increases with the priority of the event.

9.6.2.2 *ETISEO* The ETISEO protocol for event detection is similar to the ETISEO protocol for object detection discussed above. Given $N^{(e)}$ event classes, one compares the number of automatically detected events N_i for each class i against the number of ground-truth events $\tilde{N}_i$.

The mean number of true positives $E[\|\text{TP}\|]$, false positives $E[\|\text{FP}\|]$, and false negatives $E[\|\text{FN}\|]$ over the event classes are computed as

$$
E[\|\text{TP}\|] = \frac{1}{N^{(e)}} \sum_{i=1}^{N^{(e)}} \min\{N_i, \tilde{N}_i\}, \tag{9.33}
$$

$$
E[\|\text{FP}\|] = \frac{1}{N^{(e)}} \sum_{i=1}^{N^{(e)}} N_i - \min\{N_i, \tilde{N}_i\}, \tag{9.34}
$$

$$
E[\|\text{FN}\|] = \frac{1}{N^{(e)}} \sum_{i=1}^{N^{(e)}} \tilde{N}_i - \min\{N_i, \tilde{N}_i\}. \tag{9.35}
$$

The ETISEO protocol evaluates:

- the precision (Eq. 9.11)
- the recall (Eq. 9.12)
- the F-score (Eq. 9.13)

to compare the results of different algorithms where the means $E[|.|]$ replace the set cardinalities $|.|$.

Precision and *recall* are computed according to Eq. (9.11) and Eq. (9.12), where again the means $E[|.|]$ replace the set cardinalities $|.|$.

Note that these scores do not account for *event duration* and *temporal overlap*. However temporal overlap is used to associate the event estimates with ground truth data. The association is based on the overlap scores of Eqs. (9.16), (9.18) and (9.19) computed time-wise (see Figure 9.3(d)) and a conveniently choosen threshold.

9.6.2.3 *i-Lids*

The Imagery Library for Intelligent Detection Systems (i-Lids) evaluation protocol [22] for event detection compares automated results against a ground truth of the event timeline. The association between output data and ground truth is defined as follows:

- TPs: ground-truth events matched within a 10 seconds window by at least one automatically triggered alarm (i.e. multiple alarms within the validation window count as a single TP).

- FPs: automatic alarms not counted as TP and that do not fall within a 5 seconds window after another FP.

- FNs: genuine alarms not resulting in any algorithm outputs.

As shown in Figures 9.6 and 9.7, the type of events or scenarios are abandoned baggage detection, parking surveillance, doorway surveillance, sterile zone surveillance and people-tracking with multiple cameras. As mentioned earlier, although the benchmarking of the first four scenarios is centred around the detection of pre-defined events, all the tasks may require video tracking and therefore this protocol can be used to indirectly assess the performance of a tracker.

Given the association between ground truth and estimated events, the i-Lids protocol uses three scores:

- precision $\mathcal{P}$ (Eq. 9.11)

- recall $\mathcal{R}$ (Eq. 9.12)

- the weighted F-score (see Eq. 9.14). The weight β is fixed to $\sqrt{35}$ and $\sqrt{60}$ for the abandoned-baggage event and the parked-vehicle detection, respectively.

For each scenario the data corpus is split into three parts, each consisting of about 24 hours of footage (50 hours for the multiple-camera scenario). Two parts, the training and testing datasets, are publicly available. The third part, the evaluation dataset, is privately retained for independent assessment of the algorithms. The annotation of the multiple camera tracking dataset conforms to the Video Performance Evaluation Resource (ViPER) de facto standard [2, 23].

Figure 9.6 Sample frames from three scenarios from the i-Lids dataset for event detection. Top row: Abandoned-baggage detection. Middle row: parked-vehicle surveillance. Bottom row: doorway surveillance. Reproduced with permission of HOSDB.

9.7 DATASETS

This section presents an overview of datasets aimed at performance evaluation and comparison of video-tracking algorithms. The datasets are organised by application area, namely surveillance, human–computer interaction and sport analysis.

9.7.1 Surveillance

Surveillance datasets consist of a large corpus of sequences recorded for more than a decade and used in various international conferences or challenges

Figure 9.7 Sample frames from two scenarios from the i-Lids dataset for event detection and object tracking. Top row: sterile zone surveillance. Bottom row: object tracking with multiple cameras. Reproduced with permission of HOSDB.

to compare the results of video tracking or event-detection algorithms. The details of these datasets are given below:

- The *PETS-2000 dataset* [24] is composed of two sequences from a single-view outdoor scenario (Figure 9.8). The camera mounted on top of a building observes the movements of a small number of pedestrians and

Figure 9.8 Sample frames from PETS-2000 dataset aimed at benchmarking single-view person and vehicle-tracking algorithms.

DATASETS **209**

(a)   (b)

(c)   (d)

Figure 9.9 Sample frames from the PETS-2001 dataset aimed at benchmarking single and multi-view tracking algorithms. (a) and (b): Fixed camera scenario ((b) omnidirectional view). (c) and (d): Views from a moving vehicle.

vehicles. A limited amount of occlusions and constant illumination conditions facilitate the tracking task. Camera calibration information is provided together with the data.

- The *PETS-2001 dataset* [24] includes five multi-view tracking scenarios (Figure 9.9). For each scenario, two pairs of sequences (one for algorithm training and one for testing) from two synchronised cameras are provided. Four datasets have similar outdoor scenarios to those in the PETS-2000 dataset. Two cameras show different semi-overlapping views from the top of a building. In the fifth scenario the camera pair is mounted inside a car cruising on a motorway. The recordings show rear and frontal views from within the car. All the sequences are captured using standard PAL cameras and common planar lenses except in scenario 4, where an omnidirectional (360° view) lens is mounted on one of the two cameras.

- The *PETS-2002 dataset* [24] is composed of eight sequences imaging a shopping mall in challenging viewing conditions. A single camera is positioned inside a shop and the goal is to consistently track over time multiple people passing in front of a shop window (Figure 9.10). Specular

Figure 9.10   Sample frames from the PETS-2002 dataset aimed at benchmarking tracking algorithms in difficult viewing conditions. The camera is positioned behind a shop window.

reflections from the window and occlusions caused by text printed on the glass deform the appearance of the targets. Also, the targets occlude each other while passing in front of the shop.

- The *PETS-2006 dataset* [24] presents sequences recorded in a large train station. Actors simulate events that should tigger security alerts by abandoning a luggage on the platform (Figure 9.11). Four scenarios are provided, each imaged from four different views.

- The *PETS-2007 dataset* [24] is aimed at event detection in public transportation networks. Similarly to the 2006 dataset, this dataset contains multi-view images from the check-in area of an airport (Figure 9.12). Events in the dataset include abandoned luggage, thefts and subjects that remain in the field of view longer than a typical passenger.

- The Context Aware Vision using Image-based Active Recognition (CAVIAR) project [25] generated a large dataset of activities. The images are from a lobby entrance and recorded activities like walking, and entering and exiting shops. Also, a few actors simulate more complex

Figure 9.11   Sample frames from the PETS-2006 dataset aimed at benchmarking automated tracking in public places like transportation hubs. Reproduced from data provided by the ISCAPS consortium.

Figure 9.12 Sample frames from the PETS-2007 dataset aimed at benchmarking automated annotation and tracking in public places like transportation hubs. The images show two views of the check-in area of an airport. Reproduced from data provided by the UK EPSRC REASON Project.

scenarios including urban fighting and abandoning packages. Note that the resolution of the images, half Phase Alternating Line (PAL) (384×288), is fairly low with respect to more recent surveillance datasets. Figure 9.13 shows a pedestrian waking in a corridor while some customers enter and exit the shops on the left.

- The Imagery Library for Intelligent Detection Systems (i-Lids) dataset for multi-target vehicle and person tracking is composed of 25 sequences from two different outdoor video-surveillance scenarios, Broadway church scenario (BW) and Queensway scenario (QW), and present a variety of illumination conditions. Figure 9.14 shows sample frames from both scenarios. The videos have a frame size of 720×480 pixels with a frame rate of 25 Hz. The ground-truth annotation is available for 121354 frames

Figure 9.13 Sample frames depicting a target from the CAVIAR dataset, a pedestrian in a shopping mall undergoing a large scale change. The green ellipses mark sample target areas. Unmarked images from the CAVIAR dataset (project IST-2001-37540).

Figure 9.14 Sample frames from i-Lids (CLEAR-2007) dataset scenarios for video surveillance; reproduced with permission of HOSDB. (a) Scenario 1: Broadway church (BW). (b) Scenario 2: Queensway (QW). The green boxes show the positional annotation for the moving targets.

(approximately 1 hour and 21 minutes of video), divided into 50 evaluation segments. The two scenarios display common activity from pedestrians and vehicles with the addition of acted behaviours that simulate illegal parking and pedestrian interactions.

9.7.2 Human-computer interaction

In this section we cover datasets for head or hand tracking and for the analysis of meeting scenarios. The details of these datasets are given below:

Head tracking A widely used head-tracking dataset is that by Stan Birchfield.[16] The dataset is composed of 16 low-resolution (125×96) sequences recorded at 30 Hz. The sequences are the result of driving a PTZ camera with a real-time tracker and therefore present a changing background. In most of the sequences only one subject is present in the scene. Each subject performs fast movements and 360° rotations. Sample frames from the dataset are presented in Figure 9.15.

Part of the Surveillance Performance EValuation Initiative (SPEVI) dataset contains five single-person *head-tracking* sequences: *toni, toni_*

---

[16] http://www.ces.clemson.edu/~stb/research/headtracker/seq.

Figure 9.15 Each row shows sample frames from one of the 16 sequences from the Birchfield head-tracking dataset. The green ellipses mark the target areas. The targets undergo complete revolutions and partial occlusions.

change_ill, *nikola_dark*, *emilio* and *emilio_turning* (Figure 9.16). In *toni* the target moves under a constant bright illumination; in *toni_change_ill* the illumination changes from dark to bright; the sequence *nikola_dark* is constantly dark. The remaining two sequences, *emilio* and *emilio_turning*, are shot with a lower-resolution camera under a fairly constant illumination.

Three sequences of the SPEVI dataset aim at evaluating *multiple head/face-tracking* algorithms (Figure 9.17). The sequences show the upper-body parts of four persons moving in front of the cameras. The targets repeatedly occlude each other while appearing and disappearing from the field of view of the camera. The sequence *multi_face_frontal* shows frontal faces only; in *multi_face_turning* the faces are frontal and rotated; in *multi_face_fast* the targets move faster than in the previous two sequences, thus increasing the impact of motion blur on the imagery. The video has a

Figure 9.16 Sample frames depicting single head/face targets from the SPEVI dataset. The green ellipses in the first two rows mark the target areas. Top row: sequence *toni*, the head performs a complete 360° revolution. Middle row: sequence *emilio_webcam*, the head performs fast motions, it is partially occluded and undergoes large scale changes. Bottom row: sample frames from other two sequences of the same dataset that present varying illumination conditions. Left: test sequence *nikola_dark*; right: *toni_change_ill*.

frame size of 640 × 480 pixels with a frame rate of 25 Hz. A summary of the sequences composing the SPEVI dataset is presented in Table 9.2.

Hand tracking One sequence from the SPEVI dataset is aimed at testing algorithms on *hand tracking* (Figure 9.18). The video is recorded with a low-resolution webcam. The unpredictability of the hand movements and the colour similarity with other skin areas add a further degree of difficulty on top of the usual video-tracking challenges.

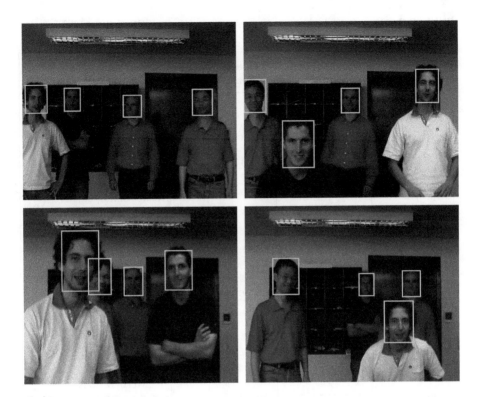

Figure 9.17 Sample frames depicting multiple head/face targets from the SPEVI dataset. The green boxes show the target positions. In these examples the four face targets occlude each other.

Meetings The *PETS-ICSV datasets* presents four indoor meeting scenarios. The goal of this data is to assess the performance of automated meeting annotation algorithms. The data for each scenario is recorded by three synchronised cameras (Figure 9.19). Two cameras with standard lenses offer top views of the meeting table. The third camera, positioned in the middle of the table, uses an omnidirectional lens. Ground-truth annotation provides information about face and eye position, gaze direction, hand gestures and facial expressions. Other meeting-scenario datasets include the Computers in the Human Interaction Loop (CHIL) corpus [26] and the Augmented Multiparty Interaction (AMI) corpus [27].

9.7.3 Sport analysis

This section covers datasets for sport analysis that include some of the PETS dataset, the CNR soccer dataset and the APIDIS basketball dataset. Due to the similar appearance of players belonging to the same

Table 9.2   Description of the sequences composing the SPEVI dataset. The frame rates are in frames per second (FPS).

Target type	Sequence	Image size	FPS	Challenges	GT
Single face/head	toni	PAL	25	Clutter self-occlusions	Yes
	toni_change_ill			Illumination changes	No
	nikola_dark			Low light	No
	emilio	320 × 240	12.5	Abrupt shifts, clutter partial occlusions	Yes
	emilio_turning			Self-occlusions	No
Multiple faces/heads	multi_face_frontal	640 × 480	25	Occlusions	Yes
	multi_face_turning			Occlusions, self-occlusions	No
	multi_face_fast			Occlusions, motion blur	No
Multi-modal	Room160	360 × 288	25	Clutter reverberations	Yes
	Room105			Occlusions, reverberations	Yes
	Chamber			Occlusions	Yes
Hand	Fast_Hand	320 × 240	12.5	motion blur, fast manoeuvres	Yes
Object	Omnicam	352 × 258	25	Lens distortion, occlusions	Yes

team, tracking algorithms may require sophisticated track management to reliably propagate the target identity across occlusions. The details of these datasets are given below.

Football The *PETS-2003* dataset consists of images from a football match recorded by three synchronised cameras. The cameras are positioned above ground level on the stands of a stadium (Figure 9.20). The ground-truth data provide information about the position of the players.

The Consiglio Nazionale delle Ricerche (CNR) soccer dataset [28] is composed of six synchronised views acquired by high definition (HD) cameras

Figure 9.18 Sample frames from the SPEVI dataset for hand tracking. The hand performs fast manoeuvres. The sequence is shot with a low-quality camera on a uniform background.

and is complemented by calibration data and two minutes of ground-truth data (Figure 9.21).

Ballsports Some test sequences of the *MPEG-4 test set* have been popular within the video-tracking community [29, 30]. Figure 9.22 shows sample frames from the MPEG-4 testbed. In this case two possible targets are the table-tennis ball and the soccer ball. Due to the sudden movements of the balls, these sequences offer interesting tracking challenges.

Basketball The Autonomous Production of Images based on Distributed and Intelligent Sensing (APIDIS) basketball dataset [31] is composed of one scenario captured from seven 2-megapixel synchronised colour cameras positioned around and on top of a basketball court. The data include camera calibration information, manual annotation of basket ball events for the entire game and one minute of manual annotation of object positions (Figure 9.23).

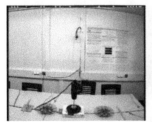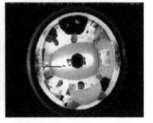

Figure 9.19 Sample frames from PETS-ICVS dataset aimed at benchmarking algorithms for automated annotation of meetings. The three images show the views from the three synchronised cameras. Reproduced from data provided by the Fgnet Project.

Figure 9.20   Sample frames from PETS-2003 dataset aimed at benchmarking automated annotation and tracking in sport footage. The three images show the different views from three synchronised cameras. Reproduced from data provided by the IST INMOVE Project.

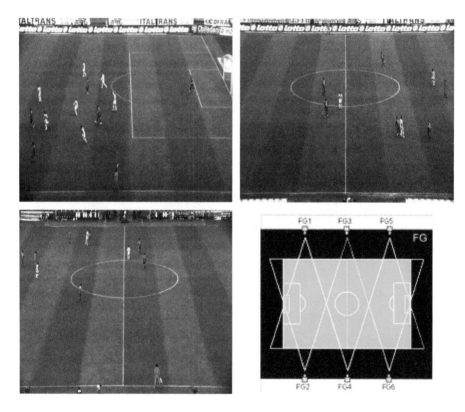

Figure 9.21   Sample frames and camera-positioning schema from the CNR soccer dataset. Reproduced with permission of ISSIA-CNR.

Figure 9.22 Sample frames depicting highly manoeuvring targets from the MPEG-4 test-set. The green ellipses mark the target areas. Top row: a ball from the *Table Tennis* test sequence. Bottom row: a football from the *Soccer* test sequence.

Figure 9.23 Sample frames from the APIDIS basketball dataset. Reproduced with permission of APIDIS Project.

9.8 SUMMARY

The development and improvement of a video-tracking algorithm requires rigorous verification of its performance in a variety of experimental conditions. This verification allows one to quantify the effects of changes in the algorithm, to compare different versions of the same algorithm when changing the value of its parameters, and to verify that the final solution is achieving the level of performance that is in line with the requirements of the application. To this end, this chapter discussed evaluation procedures for video tracking. Starting from simple quality scores like those based on the Euclidean distance, we have shown how to extend this evaluation to multi-target trackers. Moreover, we discussed how, with the help of data-association methodologies, the visual tracking problem can be treated as a binary classification problem. We also described a set of evaluation scores and protocols used in international performance-evaluation campaigns. Finally, we presented publicly available test datasets and commented on their characteristics. As the number of available datasets is ever increasing, we are keeping an updated list at www.videotracking.org

REFERENCES

1. S. Pingali and J. Segen. Performance evaluation of people tracking systems. In *IEEE Workshop on Applications of Computer Vision*, Puerto Rico, 1966, 33–38.

2. V.Y. Mariano, M. Junghye, P. Jin-Hyeong, R.Kasturi, D.Mihalcik, Li Huiping, D. Doermann and T. Drayer. Performance evaluation of object detection algorithms. In *Proceedings of the International Conference on Pattern Recognition*, Vol. 3, Quebec City, Canada, 2002, 965–969.

3. C.J. Needham and R.D. Boyle. Performance evaluation metrics and statistics for positional tracker evaluation. In *Proceedings of the International Conference on Computer Vision Systems*, Graz, Austria, 2003, 278–289.

4. J. Black, T.J. Ellis and P. Rosin. A novel method for video tracking performance evaluation. In *Joint IEEE Workshop on Visual Surveillance and Performance Evaluation of Tracking and Surveillance (VS-PETS)*, Nice, France, 2003, 125–132.

5. T. Schlogl, C. Beleznai, M. Winter and H. Bischof. Performance evaluation metrics for motion detection and tracking. In *Proceedings of the IEEE International Conference on Pattern Recognition*, Vol. 4, Hong Kong, China, 2004, 519–522.

6. D. Doermann and D. Mihalcik. Tools and techniques for video performance evaluation. In *Proceedings of the IEEE International Conference on Pattern Recognition*, Vol. 4, Barcelona, Spain, 2000, 167–170.

7. C. Jaynes, S. Webb, R.M. Steele and Q. Xiong. An open development environment for evaluation of video surveillance systems. In *Joint IEEE Workshop on Visual Surveillance and Performance Evaluation of Tracking and Surveillance (VS-PETS)*, Copenhagen, Denmark, 2002, 32–39.

8. W. Hao, A.C. Sankaranarayanan and R. Chellappa. Online empirical evaluation of tracking algorithms. *IEEE Transactions on Pattern Analysis Machine Intelligence*, 31:1443–1458, 2010.

9. J.C. San Miguel, A. Cavallaro and J.M. Martinez. Evaluation of on-line quality estimators for object tracking. In *Proceedings of the IEEE International Conference on Image Processing*, Hong Kong, 2010, 26–29.

10. J. Hoffman and R. Mahler. Multitarget miss distance and its applications. In *Proceedings of the International Conference on Information Fusion*, Vol. I, Annapolis, MD, 2002, 149–155.

11. B.N. Vo, S.R. Singh and A. Doucet. Sequential Monte Carlo implementation of the PHD filter for multi-target tracking. In *Proceedings of the International Conference on Information Fusion*, Vol. 2, Cairns, Australia, 2003, 792–799.

12. B.-T. Vo, B.-N. Vo and A. Cantoni. Analytic implementations of the cardinalized probability hypothesis density filter. *IEEE Transactions on Signal Processing*, 55(7):3553–3567, 2007.

13. H. Zhang, Z. Jing and S. Hu. Gaussian mixture cphd filter with gating technique. *Signal Processing*, 89(8):1521–1530, 2009.

14. J.E. Hopcroft and R.M. Karp. An $n^{2.5}$ algorithm for maximum matchings in bipartite graphs. *Siamese Journal of Computing*, 2(4):225–230, 1973.

15. B. Student. The probable error of a mean. *Biometrika*, 6(1):1–25, 1908.

16. M.J. Crawley. *The R Book*. New York, John Wiley & Sons, Inc., June 2007.

17. ETISEO. Evaluation du Traitement et de l'Interpretation de Sequences vidEO website. http://www-sop.inria.fr/orion/ETISEO/. Last visited: 18 March 2010.

18. R. Kasturi. *Performance evaluation protocol for face, person and vehicle detection & tracking in video analysis and content extraction*. Computer Science a Engineering University of South Florida, 2006.

19. R. Stiefelhagen, R. Bowers and J. Fiscus. Multimodal technologies for perception of humans. *Lecture Notes in Computer Science*, 4625, 2008.

20. TRECVID. Text REtrieval Conference VIDeo evaluation website. http://www-nlpir.nist.gov/projects/trecvid/.

21. F. Ziliani, S. Velastin, F. Porikli, L. Marcenaro, T. Kelliher, A. Cavallaro, and P. Bruneaut. Performance evaluation of event detection solutions: the creds experience. In *Proceedings of the IEEE Conference on Advanced Video and Signal Based Surveillance*, Como, Italy, 2005, 15–16.

22. i-Lids. Imagery Library for Intelligent Detection Systems website. www.i-lids.org.

23. ViPER. The video performance evaluation resource website. http://viper-toolkit.sourceforge.net/. Last visited: 11 October 2009.

24. PETS. Performance Evaluation of Tracking and Surveillance datasets. http://ftp.pets.rdg.ac.uk/. Last visited: 3 April 2010.

25. CAVIAR. Action recognition dataset. http://homepages.inf.ed.ac.uk/rbf/CAVIAR/. Last visited: 11 October 2009.

26. CHIL. Computers in the Human Interaction Loop website. http://www.chil.server.de. Last visited: 18 March 2010.

27. AMI. Augmented Multiparty Interaction website. http://www.amiproject.org. Last visited: 18 March 2010.

28. T. D'Orazio, M. Leo, N. Mosca, P. Spagnolo, and P.L. Mazzeo. A semi-automatic system for ground truth generation of soccer video sequences. In *Proceedings of the IEEE International Conference on Advanced Video and Signal-Based Surveillance*, Genoa, Italy, 2009, 559–564.

29. D. Comaniciu, V. Ramesh and P. Meer. Kernel-based object tracking. *IEEE Transactions on Pattern Analysis and Machine Intelligence*, 25(5):564–577, 2003.

30. E. Maggio and A. Cavallaro. Hybrid particle filter and mean shift tracker with adaptive transition model. In *Proceedings of the IEEE International Conference on Acoustics, Speech, and Signal Processing*, Vol. 2, Philadelphia, PA, 2005, 221–224.

31. APIDIS. Autonomous Production od Images based on Distributed and Intelligent Sensing dataset. http://www.apidis.org/Dataset/. Last visited: 3 April 2010.

EPILOGUE

This book offered a unified overview on the conceptual as well as implementation issues, choices and trade-offs involved in the design of video trackers. We examined various video-tracking application scenarios, covering the state-of-the-art in algorithms and discussing practical components that are necessary in application development. In particular, we highlighted the importance of understanding the image-formation process in order to effectively select the features for representing a target, based on the task at hand. Moreover, we discussed the relevance of an appropriate definition of evaluation measures to test and improve a tracking algorithm, as well as to compare results in real-world scenarios.

Throughout the book we discussed the algorithms' weaknesses and strengths, and we compared them using a set of practical and easily understandable performance measures. In addition to discussing important aspects of the design framework for video tracking, the book introduced ideas about open challenges. A non-exhaustive list of these challenges is presented below:

Target model customisation Extensive efforts have been devoted in the literature to localisation algorithms, to pre-learning a representation of the object or to learning to detect specific classes of targets. Less attention has been given to bridging the gap between generic and target-specific models. An important open challenge is now to enable a video tracker to automatically customise the target model on-line, thus reducing the burden on manually tagging the data and increasing the flexibility of the algorithms.

Tracking in crowds The performance of state-of-the-art video trackers tends to degrade with increasing number of interacting targets in the scene or simply with their density. Another open challenge is therefore the development of video-tracking algorithms that enable the analysis of behaviours in unconstrained public scenarios, such as a train station or a large square, where target might be only partially visible for long observation periods.

Track-before-detect In multi-target video tracking it may be desirable to merge the detection and the association steps into a single step where the position of multiple targets is estimated directly from the image information. Although similar algorithms performing track-before-detection have been applied successfully to radar tracking, their application to video tracking is still limited and a principled solution has to be found.

Multi-sensor fusion Multi-view geometry has been widely used to fuse point measurements from multiple cameras or from multiple frames of a moving camera. However, effective fusion of information from heterogeneous sensors (such as cameras capturing different portions of the electromagnetic spectrum), especially when loosely synchronised and aligned, is still an open area for research. The use of multiple sensors in large-scale tracking opens up challenges in the management of the computational resources and the definition of collaborative tracking strategies that have to be scalable across large systems.

We hope these challenges will be taken up by researchers and practitioners for developing novel algorithms based on the potentials of video-tracking applications in real-world scenarios.

FURTHER READING

For the reader who wishes to extend their knowledge of specific tracking problems and their current solutions, we list below other books covering various aspects of tracking. Moreover, we also provide a list of review and survey papers that cover the topic.

- Books covering *specific video-tracking applications*, such as eye tracking and camera tracking:

  - A.T. Duchowski, *Eye Tracking Methodology: Theory and Practice*, London, Springer, 2007

  - T. Dobbert, *Matchmoving: The Invisible Art of Camera Tracking*, New York, Sybex, 2005

- Books covering *specific approaches* to video tracking, such as contour-based trackers:

  - A. Blake and M. Isard, *Active Contours: The Application of Techniques from Graphics, Vision, Control Theory and Statistics to Visual Tracking of Shapes in Motion*, London, Springer, 1998

  - J. MacCormick, *Stochastic Algorithms for Visual Tracking: Probabilistic Modelling and Stochastic Algorithms for Visual Localisation and Tracking*, London, Springer, 2002

Video Tracking: Theory and Practice. Emilio Maggio and Andrea Cavallaro
© 2011 John Wiley & Sons, Ltd

- Relevant *review papers* on video tracking include:

  - A. Blake, *Visual tracking: a short research roadmap*. In *Mathematical Models of Computer Vision: The Handbook*, eds. O. Faugeras, Y. Chen and N. Paragios, London, Springer, 2005

  - A. Yilmaz, O. Javed, M. Shah, Object tracking: A survey, ACM Computing Surveys, Vol. 38(4), 13, 2006

- A *high-level mathematical perspective* of the tracking problem, in general, is given in:

  - R.P.S. Mahler, *Statistical Multisource-multitarget Information Fusion*, Norwood, MA, Artech House, 2007

  - A. Doucet, J.F.G. de Freitas and N.J. Gordon, *Sequential Monte Carlo Methods in Practice*, London, Springer, 2000

  - B.D.O. Anderson and J.B. Moore, *Optimal filtering*, Englewood Clifs, NJ, Prentice-Hall, 1979.

  - A. Gelb, *Applied optimal estimation*, Cambridge, MA, The MIT Press, 1974

  - A. H. Jazwinski, *Stochastic processes and filtering theory*, New York, NY, Accademic Press, 1970

Note that although this analysis does not extend straigtforwardly to real-world video-tracking applications, the basic principles are useful for a researcher or a practictioner interested in video tracking.

- A general overview of multi-target tracking techniques and algorithms from the *radar tracking* world can be found in the following books:

  - Y. Bar-Shalom and T.E. Fortmann, *Tracking and Data Association*, New York, Academic Press, 1988

  - S. S. Blackman, R. Popli, *Design and Analysis of Modern Tracking Systems*, Norwood, MA, Artech House, 1999

  - Y. Bar-Shalom and T. Kirubarajana, X.R. Li, *Estimation with Applications to Tracking and Navigation*, New York, John Wiley & Sons, Inc., 2002

  - B. Ristic, S. Arulampalam and N. Gordon, *Beyond the Kalman Filter: Particle Filters For Tracking Applications*, Norwood, MA, Artech House, 2004

As the treatment of the multi-target problem is limited in the video-tracking literature, principles from radar tracking can be valuable. Note that in the second book in this list, video tracking is treated as one of the many tracking applications.

- Other relevant books from the *radar tracking* literature include:

  - K. V. Ramachandra, *Kalman Filtering Techniques for Radar Tracking*, Boca Raton, FL, CRC Press, 2000

  - F. Gini and M. Rangaswamy, *Knowledge Based Radar Detection, Tracking and Classification*, New York, John Wiley & Sons, Inc., 2008

  - E. Brookner, *Tracking and Kalman Filtering Made Easy*, New York, John Wiley & Sons, Inc., 1998

  - M.O. Kolawole, *Radar Systems, Peak Detection and Tracking*, Burlington, MA, Newnes, 2002

- Finally, the *multi-camera tracking* problem is briefly covered in:

  - J. Omar and M. Shah, *Automated Multi-Camera Surveillance: Algorithms and Practice*, New York, Springer, 2008

  - Part 4 of: H. Aghajan and A. Cavallaro, *Multi-camera Networks: Principle and Applications*, New York, Academic Press, 2009

APPENDIX A

COMPARATIVE RESULTS

This Appendix presents comparative numerical results for selected methods discussed in Chapter 5 (Sections A.1 and A.2), Chapter 6 (Section A.3), Chapter 7 (Section A.4) and Chapter 8 (Section A.5) of the book. The numerical results are complemented by representative keyframes that exemplify the performance of the various video trackers and the consequences of specific algorithmic choices.

A.1 SINGLE VERSUS STRUCTURAL HISTOGRAM

A.1.1 Experimental setup

In this section we compare the results of using a single colour histogram (OH) against the results of the structural histogram (SH) representation using three localisation algorithms: mean shift (MS) (Section 5.2.1), particle filter in its CONDENSATION version (PF-C) (Section 5.3.2), and the hybrid-mean-shift-particle-filter (HY) described in Section 5.3.3.2.

Video Tracking: Theory and Practice. Emilio Maggio and Andrea Cavallaro
© 2011 John Wiley & Sons, Ltd

We select from the dataset presented in Section 9.7 six sequences containing 10 heterogeneous targets with different motion behaviours. The testbed includes sequences with:

- highly manoeuvring targets:
  - HD: one hand (Figure 9.18)
  - BT: a tennis table ball (Figure 9.22)
  - BF: a football (Figure 9.22)
- five slow pedestrians from Performance Evaluation of Tracking and Surveillance (PETS) dataset (Figure 9.9);
- $H6$: one head with background clutter (Figure 9.16, middle row)
- OF: one off-road vehicle from an aerial sequence.

To enable a fair comparison, each algorithm is initialised manually using the first target position defined by the ground-truth annotation. Colour histograms are calculated in the Red Green Blue (RGB) space quantised with $8 \times 8 \times 8$ bins. For the evaluation we use the lost-track ratio (λ, see Section 9.5.1), the overlap error ($\mathcal{D}$, Eq. 9.17) and the normalised centroid error (e, Eq. 9.7).

A.1.2   Discussion

Figure A.1 summarises the results for the three algorithms (i.e. MS, PF-C and HY) with and without the proposed multi-part target representation (SH), by displaying the average scores over the whole dataset. The complete results for each sequence together with the standard deviations σ over multiple runs are available in Table A.1.

Figure A.1 shows that MS-SH, PF-C-SH and HY-SH (the algorithms using the structural histogram representation) have better performance than their counterparts (MS-OH, PF-C-OH and HY-OH) in terms of average lost tracks (λ), shape ($\bar{\mathcal{D}}$) and centroid ($\bar{e}$) errors. Overall, the best algorithm is HY-SH. In particular, SH improves the tracking performance when a target has a non-uniform colour distribution (Table A.1). For example, HY-OH outperforms HY-SH only on the small and uniformly coloured tennis-table ball (BT). Also, due to the lower sampling of the subparts, the structural estimation of the MS vector becomes more unstable than that based on a single histogram. This problem can be solved by analysing the target and using a target-size threshold under which the single-histogram representation should be used. A few results of Table A.1 need further discussion: in HD HY-OH and HY-SH have similar λ, but HY-SH improves by around 20% in terms of $\bar{\mathcal{D}}$ and $\bar{e}$; in the easier sequences $H6$ and OF, the track is never lost, but better performance is achieved again on shape and centroid position estimation. A visual comparison is shown in Figures A.2 and A.3. Unlike HY-OH, HY-SH

Table A.1   Comparison of tracking performance between the structural representation (SH) with the single histogram (OH) for sequences with decreasing complexity (from top to bottom). Results are reported for three localisation algorithms: mean shift (MS), CONDENSATION (PF-C) and the hybrid particle-filter-mean-shift tracker (HY), Bold indicates the best result for the corresponding performance measure. Due to the deterministic nature of MS, the standard deviation on the results is presented for PF-C and HY only. Reproduced with permission of Elsevier [1].

			MS		PF-C		HY	
			OH	SH	OH	SH	OH	SH
HIGHLY MANEUVERING	HD	λ	0.46	0.44	0.30	0.25	**0.02**	0.03
		σ_λ			0.18	0.06	0.01	0.01
		$\bar{D}$	0.33	0.25	0.40	0.35	0.29	**0.23**
		$\sigma_{\bar{D}}$			0.03	0.02	0.01	0.00
		$\bar{e}$	0.30	0.24	0.35	0.32	0.30	**0.23**
		$\sigma_{\bar{e}}$			0.02	0.01	0.00	0.00
	BT	λ	0.84	0.68	0.40	0.43	**0.10**	0.14
		σ_λ			0.07	0.08	0.06	0.10
		$\bar{D}$	0.53	0.24	0.36	0.31	**0.15**	**0.15**
		$\sigma_{\bar{D}}$			0.06	0.06	0.02	0.01
		$\bar{e}$	0.37	0.19	0.28	0.26	**0.14**	0.15
		$\sigma_{\bar{e}}$			0.03	0.03	0.02	0.01
	BF	λ	0.91	0.91	0.30	0.11	0.05	**0.04**
		σ_λ			0.37	0.21	0.05	0.02
		$\bar{D}$	0.15	0.13	0.30	0.30	**0.25**	**0.25**
		$\sigma_{\bar{D}}$			0.09	0.05	0.02	0.01
		$\bar{e}$	0.18	0.16	**0.17**	**0.17**	0.19	0.19
		$\sigma_{\bar{e}}$			0.02	0.02	0.01	0.01
CLUTTER	PETS	λ	0.26	0.31	0.17	0.02	0.16	**0.01**
		σ_λ			0.04	0.02	0.06	0.00
		$\bar{D}$	0.28	0.22	0.26	**0.20**	0.25	**0.20**
		$\sigma_{\bar{D}}$			0.01	0.01	0.01	0.00
		$\bar{e}$	0.32	0.24	0.24	**0.15**	0.23	**0.15**
		$\sigma_{\bar{e}}$			0.01	0.00	0.01	0.00
	H6	λ	0.04	0.01	**0.01**	**0.01**	**0.01**	**0.01**
		σ_λ			0.00	0.00	0.00	0.00
		$\bar{D}$	0.25	0.18	0.20	00.18	0.20	**0.17**
		$\sigma_{\bar{D}}$			0.00	0.00	0.00	0.00
		$\bar{e}$	0.22	0.16	0.18	00.17	0.19	**0.16**
		$\sigma_{\bar{e}}$			0.00	0.00	0.00	0.00
EASY	OF	λ	**0.00**	**0.00**	**0.00**	0.02	**0.00**	**0.00**
		σ_λ			0.00	0.00	0.00	0.00
		$\bar{D}$	0.26	0.25	0.29	0.29	0.28	**0.25**
		$\sigma_{\bar{D}}$			0.03	0.00	0.01	0.00
		$\bar{e}$	0.27	0.20	0.20	**0.13**	0.21	0.15
		$\sigma_{\bar{e}}$			0.00	0.00	0.00	0.00

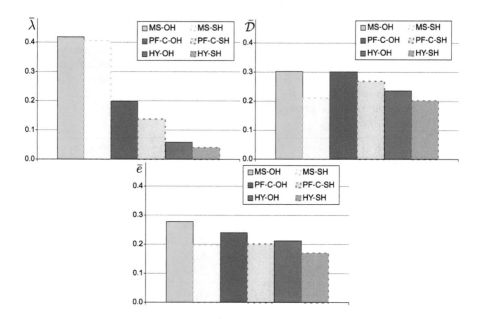

Figure A.1 Comparison of tracking results between the target representation with a single histogram (OH suffix) and the structural histogram (SH suffix). The bar plots show the average performance scores over the targets in the evaluation dataset. SH improves the tracking result for all the algorithms under analysis.

is not attracted to false targets with similar colour properties: the spatial information introduced in the model avoids a lost track and improves the overall quality (Figure A.2, left). HY-OH, using a single colour histogram, generates wrong orientation and size estimates (Figure A.3, left) as the target and the background have similar colours, and the representation is not able to correctly distinguish the face. The spatial information in HY-SH solves this problem (Figure A.3, bottom). Moreover, when a target is partially occluded (Figure A.3) the spatial information improves the final estimate.

As for the *computational complexity*, we measured the running time of the C++ implementation of the algorithms (Pentium 4, 3GHz, with 512MB of RAM). For a fair comparison, we used *H6*, where none of the trackers fails. Also, the complexity is linearly dependent on the number of pixels inside the target area (Eq. 4.4); consequently, the results for *H6*, the largest target, represent an upper bound over the dataset.

MS, PF-C and HY, using the single colour histogram representation on the 3D state space, take 1.1 ms, 6.9 ms and 5.1 ms per frame, respectively;[1] whereas using the SH representation, the values are 2.4 ms (MS-SH), 15.6 ms

---

[1] The computational time measures were obtained using the C function `clock` and do not include frame acquisition and decoding.

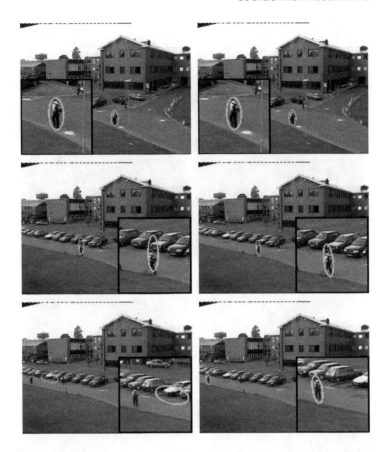

Figure A.2 Sample tracking results for the test scenario *PETS* (frames 1520, 1625 and 1758 from the sequence *Camera 1, training*) using different target representations: HY-OH (left) and HY-SH (right). Reproduced with permission of Elsevier [1].

(PF-C-SH) and 15.9 ms (HY-SH). HY-SH, MS-SH and PF-C-SH are two to three times slower than their single colour histogram counterparts. Bin selection (i.e. evaluating $b(.)$ in Eq. 4.4), the computational bottleneck, is performed only once per pixel (for the histogram of the whole ellipse) as the memory locations of the bins of the two corresponding overlapping parts (Figure 4.8) can be obtained simply by incrementing the value of the pointer computed for the whole ellipse.

A.2 LOCALISATION ALGORITHMS

A.2.1 Experimental setup

We evaluate the performance of the Hybrid-Particle-Filter-Mean-Shift tracker (HY) against MS [2] and PF-C [3] using the classical colour histograms

Figure A.3 Sample tracking results for the test sequence *H6* (frames 715, 819, and 864) using different target representations: HY-OH (left) and HY-SH (right). Reproduced with permission of Elsevier [1].

defined in Section 4.3.2. Finally Particle Filter, CONDENSATION implementation (PF-C), and Hybrid-Particle-Filter-Mean-Shift tracker (HY) are compared using 5D and 3D state models.

The state model is composed of target position, (u, v) and target size h in the 3D case. Eccentricity e and rotation are added in the 5D case.

In this case we select from the dataset presented in Section 9.7 six sequences containing 10 heterogeneous targets with different motion behaviours. The testbed includes sequences with:

- highly manoeuvering targets:
    - *HD*: one hand (Figure 9.18)
    - *BT*: a tennis table ball (Figure 9.22)
    - *BF*: a football (Figure 9.22)

- five slow pedestrians from PETS surveillance scenario (Figure 9.9)
- *H6*: one head with background clutter (Figure 9.16, middle row)
- *OF*: one off-road vehicle from an aerial sequence.

The parameter settings are described in the following:

- Colour histograms: RGB space quantised with $N_b = 10 \times 10 \times 10$ bins
- mean shift (MS) runs five times with the following kernel sizes:

  - same size as the previous frame
  - $+/-5\%$ of the size in the previous frame
  - $+/-10\%$ of the size in the previous frame.

- The Gaussian noise on the state m_{k-1} has standard deviations $\sigma_u = \sigma_v = 7$, $\sigma_h = 0.07$, $\sigma_e = 0.03$ and $\sigma_\theta = 5.0^o$. Note that the scale change is a percentage, the position is in pixels and the angle in degrees.
- PF-C uses 150 samples on the 3D state space and 250 samples on the 5D state space.
- HY uses 25% of the samples used by PF-C.
- The number of MS iterations in HY is limited to three for each particle to prevent the particles from converging on the maxima and to maintain differentiation within the sample set.

A.2.2 Discussion

Table A.2 summarises the results for the three algorithms under analysis: MS, PF-C and HY. Note that HY outperforms MS all over the dataset, except for the target *OF* ($\bar{D} = 0.26$ and 0.28 for MS and HY, respectively), thus confirming the stability of MS on targets with a limited motion. Compared with PF-C, a clear advantage of HY is when the state-transition model does not correctly predict the behaviour of the target. In our PF-C implementation using a zero-order dynamic model particles are denser around the previous state position. The faster the target, the smaller the density of the particles around it. HY eliminates this problem using the MS procedure, leading to a largely improved performance for *HD*, *BT* and *BF*. As discussed in Chapter 9, the low value of $\bar{\eta}$ returned by PF-C in *BF* has to be disregarded due to the large performance gap indicated by λ. HY shows similar performance to PF-C in tracking slow targets like those of sequence *PETS*, and targets *H6* and *OF*. The results of Table A.2, and the observation that HY uses 75% less particles than PF-C, demonstrates our claim on the improved sampling efficiency of HY compared to PF-C.

Table A.2 Comparison of tracking performance for targets with decreased complexity (from top to bottom). Bold indicates the best result for each performance score. Reproduced with permission of Elsevier [1].

			3D			5D	
			MS	PF-C	HY	PF-C	HY
HIGHLY MANOEUVRING	HD	λ	0.46	0.30	**0.02**	0.33	**0.05**
		σ_λ	0.00	0.18	0.01	0.16	0.10
		$\bar{D}$	0.33	0.40	**0.29**	0.43	**0.29**
		$\sigma_{\bar{D}}$	0.00	0.03	0.01	0.03	0.01
		$\bar{\eta}$	**0.30**	0.35	**0.30**	0.36	**0.30**
		$\sigma_{\bar{\eta}}$	0.00	0.02	0.00	0.02	0.01
	BT	λ	0.84	0.40	**0.10**	0.37	**0.08**
		σ_λ	0.00	0.07	0.06	0.08	0.02
		$\bar{D}$	0.53	0.36	**0.15**	0.31	**0.14**
		$\sigma_{\bar{D}}$	0.00	0.06	0.02	0.04	0.01
		$\bar{\eta}$	0.37	0.28	**0.14**	0.27	**0.13**
		$\sigma_{\bar{\eta}}$	0.00	0.03	0.02	0.02	0.01
	BF	λ	0.91	0.30	**0.05**	0.46	**0.04**
		σ_λ	0.00	0.37	0.05	0.42	0.02
		$\bar{D}$	0.15	0.30	**0.25**	**0.25**	**0.25**
		$\sigma_{\bar{D}}$	0.00	0.09	0.02	0.12	0.01
		$\bar{\eta}$	0.18	**0.17**	0.19	**0.15**	0.19
		$\sigma_{\bar{\eta}}$	0.00	0.02	0.01	0.03	0.01
CLUTTER	PETS	λ	0.26	**0.17**	**0.16**	0.23	**0.19**
		σ_λ	0.00	0.04	0.06	0.05	0.04
		$\bar{D}$	0.28	0.26	**0.25**	0.31	0.31
		$\sigma_{\bar{D}}$	0.00	0.01	0.01	0.02	0.02
		$\bar{\eta}$	0.32	0.24	**0.23**	0.29	0.30
		$\sigma_{\bar{\eta}}$	0.00	0.01	0.01	0.02	0.02
	H6	λ	0.04	**0.01**	**0.01**	**0.01**	0.02
		σ_λ	0.00	0.00	0.00	0.00	0.01
		$\bar{D}$	0.25	**0.20**	**0.20**	0.28	**0.24**
		$\sigma_{\bar{D}}$	0.00	0.00	0.00	0.01	0.01
		$\bar{\eta}$	0.22	**0.18**	0.19	0.19	0.19
		$\sigma_{\bar{\eta}}$	0.00	0.00	0.00	0.00	0.00
EASY	OF	λ	**0.00**	**0.00**	**0.00**	**0.00**	**0.00**
		σ_λ	0.00	0.00	0.00	0.00	0.00
		$\bar{D}$	**0.26**	0.29	0.28	**0.31**	**0.30**
		$\sigma_{\bar{D}}$	0.00	0.03	0.01	0.03	0.01
		$\bar{\eta}$	0.27	**0.20**	0.21	**0.21**	**0.21**
		$\sigma_{\bar{\eta}}$	0.00	0.00	0.00	0.00	0.00
Average		λ	0.42	0.20	**0.06**	0.23	**0.06**
		σ_λ	0.00	0.11	0.03	0.12	0.03
		$\bar{D}$	0.30	0.30	**0.24**	0.31	**0.25**
		$\sigma_{\bar{D}}$	0.00	0.03	0.01	0.04	0.01
		$\bar{\eta}$	0.28	0.24	**0.21**	0.24	**0.22**
		$\sigma_{\bar{\eta}}$	0.00	0.01	0.01	0.02	0.01

(a)   (b)   (c)

Figure A.4 Visual comparison of tracking performance. Target *BF*, frames 1, 9, 18 and 52. (a): mean shift (MS); (b): particle filter (PF-C); (c): hybrid tracker (HY). When the player kicks the ball, due to fast accelerations MS and PF-C lose the target. By combining the two algorithms HY successfully tracks the object. Reproduced with permission of Elsevier [1].

Sample results from the target *BF* are shown in Figure A.4. The target is moving in unexpected directions with shifts larger than the kernel size. Moreover, the target is affected by motion blur that decreases the effectiveness of the MS vector. HY is more stable in maintaining the track of the balls (Figure 5.12 (a)–(c)) and reduces by 75% and 83% the value of λ (see

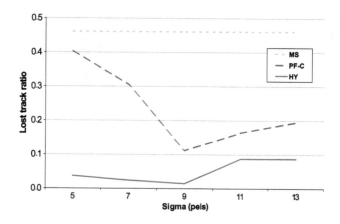

Figure A.5 Comparison of tracking results when varying the displacement random noise ($\sigma_u = \sigma_v$) of the motion model for the sequence *HD* (3D state space). It can be seen that HY is more stable than PF-C (MS is shown as a reference).

Table A.2) for the two *ball* targets, respectively. In *BT*, PF-C recovers the target after losing it, but then it fails again. In *BF*, MS and PF-C are not able to track the ball, whereas HY is fast in reacting to the abrupt shift of the ball (Figure A.4 (a)–(c)).

Figure A.5 shows the tracking results of PF-C and HY while varying the values of the displacement noise of the motion model ($\sigma_u = \sigma_v$). As a reference, the results of MS are also displayed. As the target performs abrupt and fast movements, the sampling based on the predicted prior (PF-C) is highly inefficient, whereas the particles concentrated on the peaks of the likelihood (HY) produce a better approximation. As a matter of fact, the MS procedure increases the stability of the algorithm with respect to the values of the parameters.

To test the robustness of the algorithms we run the trackers on several temporal subsampled versions of the sequence *H6*. This test simulates possible frame losses in the video acquisition phase, or an implementation of the algorithm embedded in a platform with limited computational resources (lower frame rate). The results (Figure A.6) show that HY is less affected than MS and PF-C by a frame-rate drop.

A.3 MULTI-FEATURE FUSION

A.3.1 Experimental setup

We demonstrate the adaptive multi-feature tracker on a dataset composed of 12 heterogeneous targets extracted from nine different tracking sequences:

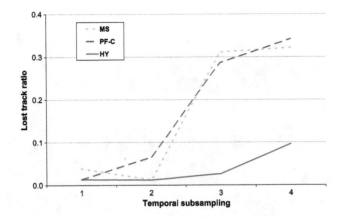

Figure A.6 Comparison of tracking results when varying the temporal subsampling rate of the input sequence with target *H6*. As the subsampling rate increases, the movement of the object becomes less predictable; HY is more stable than PF-C and MS, and achieves a lower λ (lost-track ratio).

- six head targets (*H1*, *H2*, *H3*, *H4*, *H5* and *H6*)
- five pedestrians (*P1*, *P2*, *P3*, *P4* and *P5*) and
- the toy target *O1*.

Sample frames with the targets are shown in Figure A.7.

The parameters of the tracker were set experimentally and are the same for all the targets (the only exception is the standard deviation for *H5*, as described below):

- The colour histograms are calculated in the RGB space with $N_{c,b} = 8 \times 8 \times 8$ bins.

- The orientation histograms are calculated using $N_{o,b} = 32$ bins.

- As all the head targets perform unpredictable abrupt shifts, the particle filter uses a zero-order motion model $x_k = x_{k-1} + m_k$, where the state noise m_k is a multi-variate Gaussian random variable with $\sigma_{h,k} = 0.05 \cdot h_{k-1}$, $\sigma_e = 0.021$ and $\sigma_\theta = 5^o$. $\sigma_u = \sigma_v = 5$ for all the targets except for *H5* where, due to the large size of the head, $\sigma_u = \sigma_v = 14$.

- The values of σ in Eq. (6.3) are set to $\sigma_c = 0.09$ for the colour, $\sigma_o = 0.13$ for the orientation.

- The particle filter uses 150 samples per frame.

- The time-filtering parameter τ for adaptive tracking is $\tau = 0.75$.

- A minimum of 45 particles is drawn from distribution of each feature (i.e. $V/N_f = 0.3$, with $N_f = 2$).

Figure A.7   The targets of the evaluation dataset. (From top left to bottom right) Head targets: *H1*, *H2*, *H3*, *H4*, *H5*), *H6*; pedestrians: from PETS-2001 (*P1*, *P2*, *P3* and *P4*, from CAVIAR dataset (P5); a toy bunny (*O1*). IEEE © [4].

A.3.2   Reliability scores

Figure A.8 shows a comparison between the two versions of the reliability score based on spatial uncertainty proposed in Section 6.3.3 (i.e. α^4 and α^5) with the existing reliability scores α^1, α^2 and α^3 obtained by applying temporal smoothing (i.e. Eq. 6.17) to Eq. (6.9) and Eq. (6.10).

By comparing the results on the complete dataset, the five scores achieve similar performances on *H5*, *H6* and *P4*, and present large error differences in sequences with occlusions and clutter (*H3*, *H4*, *P1*, *P2* and *P3*). In particular, α^1 and α^3 leads to poor results compared to α^2, α^4 and α^5 in *H3*, *H4*, *P1*, *P2* and *P3*. The two scores with faster variability (Figure 6.3) are also less accurate, especially in sequences with false targets and clutter. In fact, the more conservative score α^2 results in a more stable performance on the same targets.

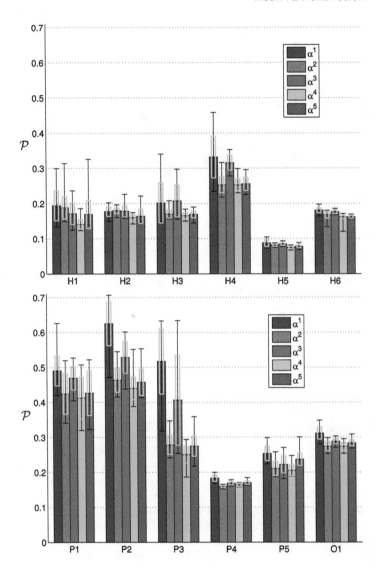

Figure A.8 Tracking results for different feature weighting strategies. The bars represent the average error, the boxes represent the standard deviation of the error and the error bars represent the maximal and minimal error on each sequence. IEEE © [4].

The score α^4 consistently yields to the lowest error and lowest standard deviation across different target types and tracking challenges. In particular, a large performance gap is achieved on *H1*, *P2* and *P3*. This confirms the observation drawn from Figure 6.3 described in Section 6.3.4: the score α^5 is more accurate than α^1, α^2 and α^3, but performs worse than α^4. For this

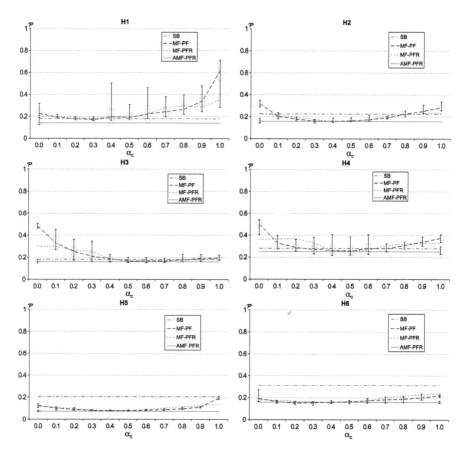

Figure A.9 Comparison of head-tracking results for AMF-PFR, different fixed combinations of the features with (MF-PFR) and without (MF-PF) multi-feature re-sampling and the head tracker (SB) proposed in [5]. The average distance from the ground truth (lines), the standard deviation (boxes) and the maximal and minimal errors (error bars) are plotted against the weight given to the colour feature. For readability purposes the error bars and standard deviations of the non-adaptive algorithms are displayed only for MF-PF (comparable results are obtained with MF-PFR). IEEE © [4].

reason, α^4 will be used in the following comparative tests as feature reliability score in the adaptive multi-feature particle filter (AMF-PFR) adopting the re-sampling procedure described in Section 6.3.

A.3.3 Adaptive versus non-adaptive tracker

Figure A.9 and Figure A.10 show the performance comparison between:

- AMF-PFR: the *adaptive* tracker

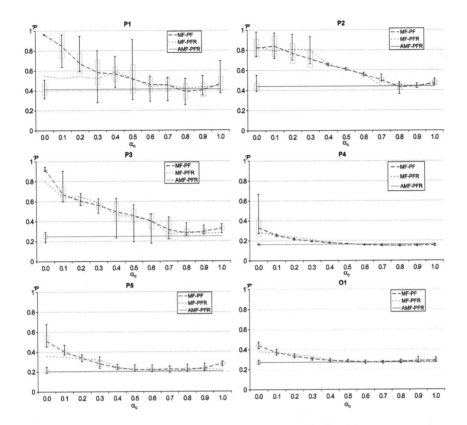

Figure A.10 Comparison of tracking results for AMF-PFR and different fixed combinations of the features with (MF-PFR) and without (MF-PF) multi-feature re-sampling. The average distance from the ground truth (lines), the standard deviation (boxes) and the maximal and minimal errors (error bars) are plotted against the weight given to the colour feature. For readability purposes the error bars and standard deviations of the non-adaptive algorithms are displayed only for MF-PF (comparable results are obtained with MF-PFR). IEEE © [4].

- MF-PFR: the tracker with multi-feature re-sampling
- MF-PF: the tracker without multi-feature re-sampling
- SB: the elliptic tracker proposed by Birchfield [5].

The results related to MF-PFR and MF-PF are obtained by fixing a priori the colour importance α_c. Note the large performance improvements when moving from single feature (i.e. MF-PF with $\alpha_c = 0$ and $\alpha_c = 1$) to multi-feature algorithms. It is worth noticing that the optimal working point, $\hat{\alpha}_c$, of the non-adaptive trackers (MF-PF, MF-PFR) varies from target to target.

For example $\hat{\alpha}_c = 0.3$ in *H1*, while $\hat{\alpha}_c = 0.6$ in *H3*. In these two cases MF-PF and MF-PFR require manual tuning to achieve optimal performance, whereas in AMF-PFR the adaptation is automated. Furthermore the error of AMF-PFR is comparable with or lower than the best result of the non-adaptive algorithms (MF-PF and MF-PFR).

On *H1* and *H5* the error of AMF-PFR is 17% and 5%, respectively, lower than the error at the best working point of MF-PFR. Figure A.12 shows sample frames on *H1* from the run with the closest error to the average: MF-PF (left column) and AMF-PFR (right column). As the target changes scale (the first two rows of Figure A.12 show a 1/4 of an octave scale change) the scale of the filter is adapted. The numbers superimposed on the images are the minimum and maximum of the standard deviation of the Gaussian derivative filters that are used to generate the scale space. When the head turns the adaptive weighting reduces the tracking error. On *H6*, although AMF-PFR is more accurate than most of the fixed combinations of weights, its performance is 5% worse than the best non-adaptive (manually set) result. In this sequence scale changes, illumination changes and a partial occlusion occur; both colour and orientation models are unable to correctly describe the target, and this results in a suboptimal adaptation of α. In this case, a larger pool of features could help to improve the effectiveness of the adaptation.

The results on *P1*, *P2*, *P3*, *P4* and *P5* (Figure A.10) show how the algorithm adapts when one feature is more informative. The orientation histogram of a walking person varies over time because of the swinging of the arms and movement of the legs; hence the orientation histogram does not contribute significantly to improving the tracker performance. By allowing time adaptation, AMF-PFR *automatically* achieves a performance that is similar to the best fixed combination of the features. Figure A.13 (left column) shows sample results on *P4*. AMF-PFR tracks the target despite the presence of clutter with similar colours (the white car) and despite the occlusion generated by the lamp post. Similarly for *O1* (Figure A.13, right column): the occlusion generated by the hand is overcome thanks to the multiple hypothesis generated by particle filter (PF) and to the adaptiveness of the target representation.

Finally, the standard deviations and error bars in Figures A.9 and A.10 show that the error is more stable for AMF-PFR than for the non-adaptive counterpart MF-PF. This is more evident in *H1*, *H4* and *P3*, where a small variation of the weights results in MF-PF losing the track. For example, Figure A.11 shows the results of the *worst run* in terms of error on target *H4*: MF-PF (left column) is attracted by false targets, and the track is lost during the rotation of the head. Although AMF-PFR (right column) does not accurately estimate the target size, the object is continuously tracked. Moreover, Figure A.9 shows also that AMF-PFR outperforms SB on all the six head targets.

Figure A.11 Sample results from the *worst run* on target *H4* (frames 116, 118, 126 and 136). Left column: non-adaptive multi-feature tracking (MF-PF). The colour contribution is fixed to 0.5, i.e. the value that gives the best average result on MF-PF. Right column: adaptive multi-feature tracker (AMF-PFR). Note that MF-PF is attracted to a false target, while the proposed method (*worst run*) is still on target. IEEE © [4].

Figure A.12 Sample results from the *average run* on target *H1* (frames 82, 92, 100, 109, 419, 429, 435 and 442). Left column: non-adaptive multi-feature tracker (MF-PF). Right column: adaptive multi-feature tracker (AMF-PFR). When the target appearance changes, AMF-PFR achieves reduced tracking error by varying the importance of the features over time. IEEE © [4].

Figure A.13 Sample results of the adaptive multi-feature tracker (AMF-PFR). Left column: target *P4* (frames 38, 181, 273 and 394). Right column: target *O1* (frames 332, 368, 393 and 402). Due to the adaptive combination of colour and orientation information and to the multiple-tracking hypothesis generated by the particle filter the tracker can successfully overcome background clutter (left column) and occlusions (right column). IEEE © [4].

Figure A.14 shows how the information encoded in the simple representation used by SB (the edges sampled on the border of the object) can be undermined by the edge-filter responses generated by the bookshelf in the background. The performance improvement is attributable to the representation of the gradient based on the orientation histograms used in AMF-PFR, which encode internal edge information that is less likely affected by clutter.

A.3.4 Computational complexity

In terms of computational complexity the adaptive tracker based on PF runs on a Pentium 4 3 GHz, at 13.2 fps on *H1* (average area: 1540 pixels), 7.4 fps on *P4* (2794 pixels) and 2.2 fps on *H5* (10361 pixels). The computational complexity approximately grows linearly with the target area. It is also worth noticing that the complexity does not depend on the frame size, as the processing is done on a region of interest around the target. AMF-PFR spends 60.3% of the time computing the orientation histograms, 31.1% on the colour histogram, 7.9% on the recursive propagation of the particles and only 0.7% computing the feature reliability scores. Also, the computational cost associated to the orientation histogram could be reduced by using an optimised implementation of the Gaussian scalespace computations [6].

A.4 PHD FILTER

A.4.1 Experimental setup

In this section we report on tests on real-world outdoor surveillance scenarios of the multi-target tracking framework that uses the particle PHD filter. The objective performance evaluation follows the VACE protocol [7]. A comprehensive description of the protocol is available in Section 9.6.1.2.

The parameters used in the simulations are the same for all test sequences and, unless otherwise stated, they are the same for both face and change detectors. The values of the parameters are empirically chosen and a sensitivity analysis for these choices is given later in the discussion section. The particle Probability Hypothesis Density (PHD) filter uses $\rho = 2000$ particles per target and $\tau = 500$ particles per detection. The standard deviations of the dynamic model defining target acceleration and scale changes are: $\sigma_{n(u)} = \sigma_{n(v)} = \sigma_{n(w)} = \sigma_{n(h)} = 0.04$. The standard deviations of the Gaussian observation noise are: $\sigma_{m(w)} = \sigma_{m(h)} = 0.15$ for the change detector and 0.1 for the face detector. Larger spatial noise is used in the change-detector case as we have to cope with the errors related to merging and splitting of the blobs. The birth-intensity parameter defining the number of new targets per frame is $\bar{s} = 0.005$. The number of observations due to clutter is set to $\bar{r} = 2.0$ clutter points per frame. The missing detection probability $p_{\mathrm{M}} = 0.05$ and the survival

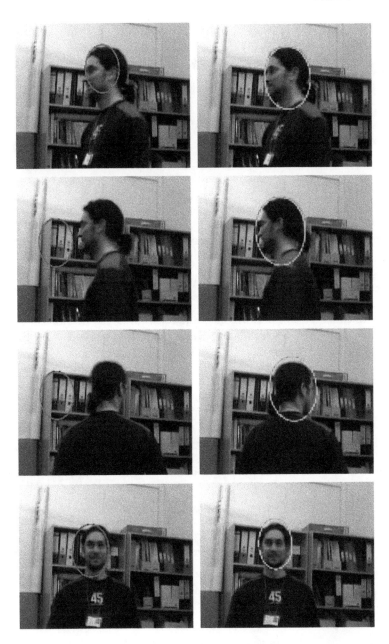

Figure A.14 Sample tracking results on target *H5* (frames 200, 204, 218 and 247). Left column: elliptic head tracker (SB). Right column: adaptive multi-feature tracker (AMF-PFR). Unlike SB, the gradient information used in the AMF-PFR target model manages to separate the target from the clutter. IEEE © [4].

probability $e_{k|k-1} = 0.995$. The new-born particles are spread around the detections with $\sigma_{b,u} = \sigma_{b,v} = \sigma_{b,w} = \sigma_{b,h} = 0.02$ and $\sigma_{b,\dot{u}} = \sigma_{b,\dot{v}} = 0.05$. The resampling strategy uses $N_s = 7$ stages. The number of resampled particles for GMM clustering is $\rho_{GM} = 500$ per target. Clusters with weight lower than $H = 10^{-3}$ are discarded, while $T_M = 0.5$ is used to accept the cluster centres as real targets. For data association, the depth of the graph is $W = 50$ and means that the algorithm is capable of resolving occlusions for a maximum of 2 seconds with a 25 Hz frame rate.

A.4.2 Discussion

Table A.3 shows the performance comparison between the SDV dynamic and observation models described in Section 7.6.1, and linear models with fixed variances. Fixing the variances is equivalent to removing from Eq. (7.23) and Eq. (7.24) all references to target width w and height h. The fixed values of the standard deviations are chosen as a trade-off between large and small targets $(\sigma_{n^{(u)}} = \sigma_{n^{(v)}} = \sigma_{n^{(w)}} = \sigma_{n^{(h)}} = 3$ and $\sigma_{m^{(w)}} = \sigma_{m^{(h)}} = 5)$. The tracker with SDV models is better in terms of both precision and accuracy. Also, the significance of the performance difference over the evaluation segments is always below the 5% validation threshold. The trade-off selected for the

Table A.3 Performance comparison between the dynamic and observation models with State Dependent Variances (SDV) and the models with Constant Variances (CV). The two testing scenarios BW (Broadway church) and QW (Queensway) are from the CLEAR-2007 dataset. The figures represent the precision (MODP and MOTP) and the accuracy (MODA and MOTA) of the two approaches. IEEE © [8]

		BW		QW	
		SDV	CV	SDV	CV
MODP	Avg	0.537	0.530	0.382	0.377
	Significance	5.55E-09		1.54E-02	
MODA	Avg	0.444	0.429	0.211	0.153
	Significance	2.63E-04		7.33E-06	
MOTP	Avg	0.544	0.536	0.388	0.381
	Significance	9.75E-08		3.37E-03	
MOTA	Avg	0.436	0.415	0.194	0.128
	Significance	2.11E-06		1.28E-06	

standard-deviation values is not appropriate near the extrema of the target scale range. When a large object (i.e. 200 pixels wide) is partially detected, the error associated with the observation z_k may be several times larger than the standard deviation. Similarly, while an acceleration of three pixels per frame may be appropriate for a middle-size target, this value is large compared to the typical motion of a pedestrian located in the camera's far-field.

To quantify the change in performance when adding the PHD filter to the tracking pipeline, we compare PHD-MT with MT, a tracker that performs data association directly on the raw detections. Figure A.15 shows the difference in terms of evaluation scores between Multiple Target tracker based on the PHD filter (PHD-MT) and Multiple Target tracker (MT). The last set of bars in the two plots shows the average results over the segments. It can be seen that the filtering of clutter and noise consistently improves both accuracy and precision for all the evaluation segments in both scenarios. In the video segments with higher levels of clutter and where tracking is more challenging, the performance improvement is larger. Similar considerations can be drawn by comparing the results of the two different scenarios. More false-positive detections are generated by the change detector with QW; by removing these false positives, the PHD-MT obtains larger improvements in terms of evaluation scores than with BW (Figure A.15).

Figure A.16 shows *accuracy and precision* scores when we change the set-up of the PHD filter parameters. Each plot was obtained by changing with a $\log_2$ scale one parameter at a time while fixing the rest to values defined at the beginning of this section. It is interesting to observe that large variations in tracking performance are associated with changes in the observation and dynamic model configuration (Figure A.16 for $\sigma_{v(.)}$ and $\sigma_{n(.)}$). Too large or too small noise variances result in insufficient or excessive filtering and produce a drop in tracking accuracy. Also, decreasing ρ (i.e. the number of particles per estimated target) reduces the quality of the filtering result as the approximation of the PHD propagation becomes less accurate. The PHD filter is less sensitive to variation of the other parameters. Averaging over a large number of targets produces a compensation effect where different parameter values are optimal for different observation and target behaviours, thus returning similar scores. Also, in the case of birth and clutter parameters ($\bar{s}$ and $\bar{r}$), low variability is associated with the fact that birth and clutter events are relatively sparse in the state and observation spaces. When varying $\bar{r}$, the average number of clutter points per scan, the result is stable until $\bar{r}$ is grossly overestimated. Similarly, only a small impact is associated with variations of missing detection (p_{M}) and survival ($e_{k|k-1}$) probabilities.

A.4.3 Failure modalities

The PHD filter was incorporated in an end-to-end flexible tracking framework that can deal with any detectors that generate a set of observations representing position and size of a target. First, clutter and spatial noise are

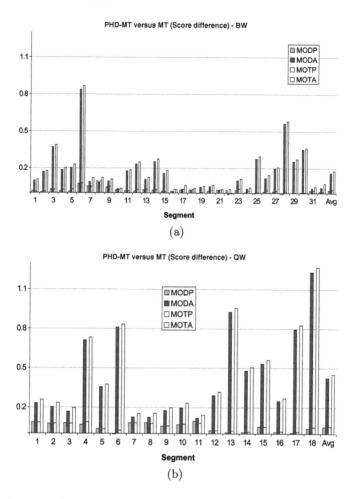

Figure A.15 Difference of tracking results between the multi-target tracker with (PHD-MT) and without (MT) PHD filter. The bar plots show the evaluation score difference between the two algorithms for all the evaluation segments in the two scenarios of the CLEAR-2007 dataset. (a): BW (Broadway church); (b): QW (Queensway). The last set of four bars shows the average difference over the segments of each scenario. Positive values correspond to performance improvements achieved with the PHD filter. IEEE © [8].

filtered by the particle PHD filter. Next, clustering is used on the samples of the PHD to detect filtered target positions. Then, the cluster centres are processed by a data-association algorithm based on the maximum path cover of a bi-partitioned graph.

The results on face, pedestrian and vehicle tracking demonstrate that it is possible to use the PHD filter to improve accuracy and precision of a

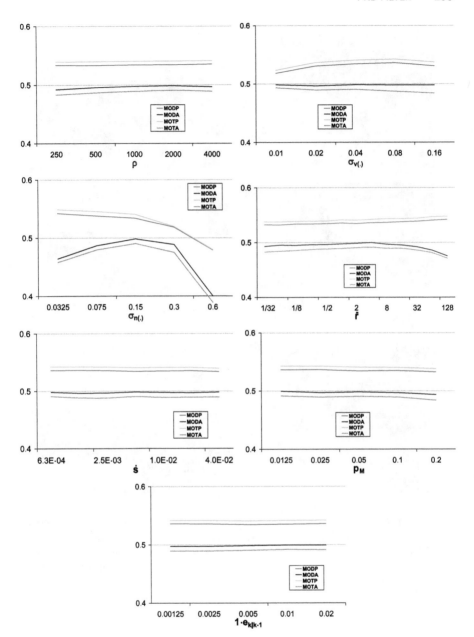

Figure A.16   Sensitivity analysis on the parameters of the PHD filter. Each plot shows the average accuracy and precision scores on the Broadway church scenario (BW) while varying one of the parameters (see Sections 7.5 and 7.6 for the definition of the parameters). IEEE © [8].

Figure A.17 Failure modalities of the particle PHD filter when using a change detector. The red boxes are the observations and the green boxes are the output of the PHD filter. Left column: inconsistent detections in the far field are interpreted by the PHD filter as clutter and therefore removed. Right column: interaction between targets (object merging) generates a bounding box for a group of objects. IEEE © [8].

multi-target tracking framework. However the quality of the filtering result depends on how much the input detections match the assumptions necessary to formulate the PHD recursion. Figure A.17 shows two examples of *failure modalities* of the particle PHD filter. The close-up images in Figure A.17 (left column) show a first failure modality. The change detector generates for the person in the far-field detections that are inconsistent over time. These

detections are considered by the PHD filter as clutter and therefore eliminated. Figure A.17 (right column) shows a sample result when one of the
assumptions of the PHD filter is violated (Section 7.5), i.e. the targets generate dependent observations. As the targets overlap, the change detector merges
the two blobs and produces one observation only. In this case the change of
size is outside the range of changes modelled as noise. When the targets split,
the delay introduced by the PHD filter generates a set of missing detections.

While splitting could be partially handled by enabling spawning from targets (Eq. 7.18), merging of observations poses a problem as the PHD was
originally designed to track using punctual observations just as for those generated in a radar scenario, where target interaction is weak. These problems
can be overcome by using a trained object detector (e.g. a vehicle detector),
within the same framework.

A.4.4   Computational cost

The *computational cost* of the particle PHD filter is comparable to that of
the two object detectors (Figure A.18). The graph-based data association
has low influence on the overall cost as the computation is based on positional information only. If more complex gain functions are used to weight
the edges of the graph (for example by comparing target appearances using
colour histograms), then the data association would significantly contribute
to the overall computational cost. The larger resource share claimed by the
particle PHD filter with the change detector, compared to the face-tracking
case, is mainly due to the larger average number of targets in the scene.

Figure A.19 shows the processing time versus the number of targets
estimated with the BW scenario. The processing time of the full tracker
(PHD-MT) is compared with that of the recursive filtering step
(PHD&GMM). The results are obtained with a non-optimised C++ implementation running on a Pentium IV 3.2 GHz. As the number of particles grows
linearly with the number of targets and the number of observations, the theoretical computational cost is also linear. The mild non-linearity of the curve

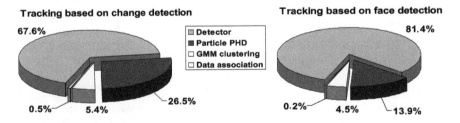

Figure A.18 Percentage of computational resources allocated to each block
of the tracker. The PHD filter requires fewer resources than the detectors.
IEEE © [8].

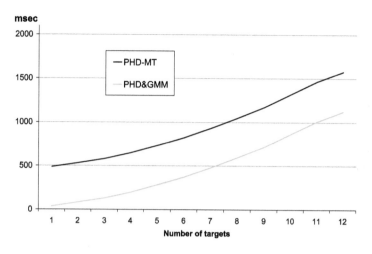

Figure A.19   Processing time versus estimated number of targets in the scene on a sequence from the CLEAR-2007 dataset. PHD-MT: full tracker; PHD&GMM: PHD filtering and GMM particle clustering steps. IEEE © [8].

PHD&GMM is due to the fact that with a low number of particles the processor performs most of the operations using the cache memory. When the number of targets increases, the filter propagates more particles and the curves become steeper as the cost is now associated with the use of off-chip memory. Also, a larger overhead for PHD-MT is due to the non-optimal implementation of the object detector (0.5 seconds/frame), and not to the filter itself. Furthermore, as most of the calculations necessary to propagate a particle depend on its previous state only, the particle PHD is well suited for a parallel implementation.

With an optimised implementation of the detector and a GPU (graphics processing unit) or multi-core implementation of the PHD filter, the tracker could achieve real-time performance.

It is of interest also to compare the computational time of PHD&GMM with the hypothetical results of a particle implementation that propagates the full multi-target posterior (FP). When one target only is visible, then the PHD and the FP resort to the same algorithm (that takes 40 milliseconds/frame). With multiple targets, because the dimensionality of the state space in FP grows, an exponential number of particles is necessary to achieve a constant density sampling. The computational time per frame of an FP implementation would then be: 1.5 seconds for two targets, 40 minutes for four targets and 187 years with eight targets. In this case, the only feasible approach would be to use a more efficient sampling method in an MCMC fashion [9]. Unlike FP, the PHD filter limits the propagation of the particles to the single target state space and thus achieves linear complexity.

A.5 CONTEXT MODELLING

A.5.1 Experimental setup

In this section we demonstrate the contribution to video tracking given by learning clutter and birth density with mixture models using a detector based on background subtraction [10]. We assess the performance using a colour statistical change detector [10]. The tests are conducted on the scenarios from the CLEAR-2007 dataset. In scenario BW we have selected a subset of sequences (dataset code: 101) as the others present slightly different camera views.

We train the models on frame spans of the sequences where ground-truth data used for testing is not available. We first run the tracker on the training data with uniform clutter and birth intensities. We then extract birth and clutter samples from the tracker output and use these samples to estimate birth and clutter intensity models.

We compare the results of a tracker that uses context-based models of clutter and birth intensity (CM) with the results obtained with uniform models of clutter and birth intensity (UM).

The intensity magnitudes of uniform models of clutter and birth intensity (UM) ($\bar{r}$ and $\bar{s}$) are the same as in the learning phase. We also compare these two solutions with six other algorithms obtained by combining different clutter and birth learning strategies:

- A1: GMM birth and uniform clutter intensities

- A2: uniform birth and GMM clutter

- A3: uniform birth and clutter, but magnitudes $\bar{r}$ and $\bar{s}$ estimated from the data

- A4: clutter as in A3 and GMM birth

- A5: birth as in A3 and GMM clutter

- A6: GMM birth and clutter intensities, but with birth interactive data collection, performed as for the clutter data.

A.5.2 Discussion

Figure A.20 shows sample results of CM with the QW scenario, where contextual feedback improves the PHD filter performance. As low birth intensity (i.e. strong temporal filtering) is estimated over the parking areas (Figure A.20), false detections on the number plate are consistently removed (Figure A.20). Compare these results with those of UM in Figure 8.3.

Figure A.21 shows a comparison of the CM and UM filtering results on scenario QW. The detections corresponding to waving branches are filtered out for a longer number of frames due to the feedback from the GMM

Figure A.20   Filtering results of the particle PHD filter using learned clutter and birth intensities (CM) on the same data as Figure 8.3(a) and (b). First row: tracker output. Second row: the detections from a background subtraction algorithm are colour-coded in red and the PHD output is colour-coded in green. Strong filtering in a parking background area (left of the image) removes the false detections on the number plate and prevents a false track.

clutter model (Figure A.21, second and third rows). Low clutter levels instead are assigned to the platform regions, thus allowing the PHD filter to validate after a few frames the coherent detections corresponding to a pedestrian (Figure A.21, first and second rows).

Figure A.22 compares the tracking results for the two scenarios from the CLEAR-2007 dataset. The values of the bars are the percentage score differences with respect to the base-line tracker UM. In both scenarios the Gaussian mixtures used to model birth and clutter intensities (CM and A6) outperform the other models, especially in terms of accuracy. CM and A6 improve the clutter-removal capabilities of the PHD filter, thus reducing false detections and false tracks. This is also confirmed by the results in Figure A.23, obtained by varying the values of clutter and birth magnitudes $\bar{r}$ and $\bar{s}$ in UM. In all cases CM outperforms UM in terms of accuracy. *This is true also for the precision scores, except when the clutter intensity is overestimated.* However, in this case UM achieves slightly better precision than CM, but at the cost of a large drop in accuracy (Figure A.23 (d)).

It is important to note that the curves produced by UM are stable around their maximum values, as by changing $\bar{r}$ and $\bar{s}$ the filtering behaviour becomes

Figure A.21   Comparison of filtering results for the BW scenario. Left column: tracker that uses learned clutter and birth intensities (CM). Right column: tracker that uses uniform intensities (UM). False detections due to waving trees are more consistently removed by using the Gaussian-mixture-based birth and clutter models (red: detections; green: PHD filter output). IEEE © [11].

more suitable on one subset of targets but suboptimal on another. This leads to similar performance scores. Also, the results in Figure A.22 show that, given the same average intensity, the GMM density estimates improve the performance with respect to the uniform distributions (compare CM with A3). Both clutter and birth intensity models contribute to the final performance improvement. However, clutter intensity trained with manually labelled data achieves better results than CM birth intensity trained using the output of the tracker (compare A1 with A2, or A3 with A4). This is due to the fact that *the birth model must account also for track reinitialisations*; the volume of the state space where a birth event is likely to happen is larger and thus the model is less discriminative than that for clutter. However, a more precise birth model trained with manually annotated data (A6) leads to ambiguous results (compare A6 and CM). On the one hand, when most false detections are generated by background clutter, as in the QW scenario (Figure A.22 (b)), a

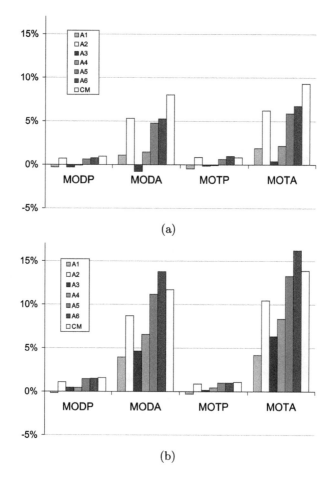

(a)

(b)

Figure A.22   Comparison of tracking accuracy and precision on the CLEAR-2007 scenarios BW (a) and QW (b). The bars represent the score percentage difference (i.e. performance improvements) with respect to the baseline algorithm using uniform birth and clutter models (UM). Each colour corresponds to an algorithm with a different level of context awareness. CM: algorithm using the complete context model. See the text for more details on the other algorithms (i.e. A1–A6). IEEE © [11].

stronger birth constraint allows A6 to outperform CM in terms of accuracy. On the other hand, when a large percentage of tracking errors is due to occlusions and blob merging (as in the BW scenario), the same constraint prevents a prompt reinitialisation of the tracks (Figure A.22 (a)).

The scores also show that *the accuracy improvement* in BW (Figure A.22 (a)) is larger than in QW (Figure A.22 (b)). Although QW is more challenging than BW, as shown by the lower scores, most tracking errors in

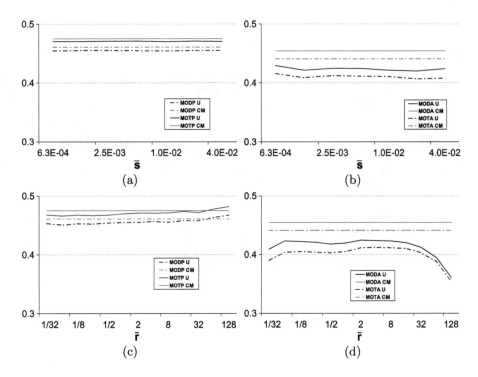

Figure A.23   Comparison of tracking performance when varying the birth and clutter magnitudes ($\bar{r}$ and $\bar{s}$) between the tracker with Gaussian-Mixture-based birth and clutter intensities (CM) and the tracker with uniform distributions (UM). (a) and (c): Scenario BW. (b) and (d): Scenario QW. IEEE © [11].

QW are not due to background clutter. In most cases, proximity between targets produces several target mergings and splittings, thus reducing the temporal consistency across consecutive detections. In the BW scenario instead, the motion of the branches generates a large number of false detections that are consistently localised. By introducing contexual information, the model adjusts the latency (i.e. the number of consecutive detections) necessary for a target to be validated.

REFERENCES

1. E. Maggio and A. Cavallaro. Accurate appearance-based bayesian tracking for maneuvering targets. *Computer Vision and Image Understanding*, 113(4):544–555, 2009.

2. D. Comaniciu, V. Ramesh and P. Meer. Kernel-based object tracking. *IEEE Transactions on Pattern Analysis and Machine Intelligence*, 25(5):564–577, 2003.

3. P. Perez, C. Hue, J. Vermaak and M. Gangnet. Color-based probabilistic tracking. In *Proceedings of the European Conference on Computer Vision*, Vol. 1, Copenhagen, Denmark, 2002, 661–675.

4. E. Maggio, F. Smeraldi and A. Cavallaro. Adaptive multifeature tracking in a particle filtering framework. *IEEE Transactions on Circuits Systems and Video Technology*, 17(10):1348–1359, 2007.

5. S. Birchfield. Elliptical head tracking using intensity gradients and color histograms. In *Proceedings of the IEEE Conference on Computer Vision and Pattern Recognition*, Santa Barbara, CA, 1998, 232–237.

6. D.G. Lowe. Object recognition from local scale-invariant features. In *Proceedings of the International Conference on Computer Vision*, Corfu, Greece, 1999, 1150–1157.

7. R. Kasturi. *Performance evaluation protocol for face, person and vehicle detection & tracking in video analysis and content extraction*. Computer Science and Engineering University of South Florida, 2006.

8. E. Maggio, M. Taj and A. Cavallaro. Efficient multi-target visual tracking using random finite sets. *IEEE Transactions on Circuits Systems and Video Technology*, Vol. 18, No. 8, pp. 1016–1027, Aug. 2008.

9. Z. Khan, T. Balch and F. Dellaert. An MCMC-based particle filter for tracking multiple interacting targets. In *Proceedings of the European Conference on Computer Vision*, Prague, Czech Republic, 2004, 279–290.

10. A. Cavallaro and T. Ebrahimi. Interaction between high-level and low-level image analysis for semantic video object extraction. *EURASIP Journal on Applied Signal Processing*, 6:786–797, 2004.

11. E. Maggio and A. Cavallaro. Learning scene context for multiple object tracking. *IEEE Transactions on Image Processing*, 18(8):1873–1884, 2009.

INDEX

www.ingramcontent.com/pod-product-compliance
Lightning Source LLC
Chambersburg PA
CBHW072109250125
20788CB00003B/15